EXPLANATION:
AN INTRODUCTION TO THE
PHILOSOPHY OF SCIENCE

Gerald Bakker
Len Clark
Earlham College

Mayfield Publishing Company
Mountain View, California

To Barbara and Mary Jo

Copyright © 1988 by Mayfield Publishing Company

All rights reserved. No portion of this book may be reproduced in any form or by any means without written permission of the publisher.

Library of Congress Cataloging-in-Publication Data

Bakker, Gerald.
    Explanation : an introduction to the philosophy of science / Gerald Bakker, Len Clark.
        p.   cm.
    Bibliography: p.
    Includes index.
    ISBN 0-87484-838-5
    1. Science—Philosophy. I. Clark, Len. II. Title.
Q175.B168   1988
501—dc19   87-31538
                                                                        CIP

Manufactured in the United States of America

10 9 8 7 6 5 4 3 2 1

Mayfield Publishing Company
1240 Villa St.
Mountain View, California 94041

Sponsoring editor, James Bull; manuscript editor, Marie Enders; cover designer, Andrew Ogus; cover illustration, Leonardo da Vinci, *Flying Machine with a Man Flying It*, (c. 1488), courtesy of Foundations de l'Institut de France. The text was set in 10/12 Goudy Oldstyle by BookMasters and printed on 50# Maple Opaque by Maple Vail.

Additional credits appear on a continuation of the copyright page, p. 210.

"Although human subtlety makes a variety of inventions . . . it will never devise an invention more beautiful, more simple, or more direct than does nature, because in her inventions nothing is lacking and nothing is superfluous."

—Leonardo da Vinci

CONTENTS

PREFACE

Origins and Character of the Text

Readers who wonder whether the philosophy of science is worth their time and effort should go directly to chapter one to begin their exploration. Others who know something about the philosophy of science will want to know why this work is organized the way it is and what we are attempting to accomplish with it.

Explanation grew out of our team teaching a course in the philosophy of science at Earlham College. It is meant to remedy a common problem of philosophy texts that introduce the most abstract subjects without having any content language on which to build and with which to capture the students' interest. It also avoids the detail and specialization of many other philosophy of science texts that render them usable and interesting primarily for philosophy students. We use the language of science to introduce students to philosophy and the philosophy of science.

To exhibit the relevance of philosophy to natural science, we focus on the concept of explanation. Explanation and the question of how scientists arrive at the best explanations provide unifying themes for the treatment of many of the great issues in the philosophy of science. Considering whether explanation is at heart causal, for example, allows an introduction to Aristotle. The consideration of explanation as merely description allows an intelligible entry to Hume, Mach, and positivism. Once such authors as Hume are introduced in the appropriate place, it is natural to follow up with a discussion of Kant or whatever other option the history of philosophy provides in response to a given position.

The text is designed to encourage students to develop their own philosophy of science and their own way of understanding explanation. Our belief is that this will allow them to become more reflective and better scientists and more thoughtful actors on the issues relating science and society. Our aim is not to convince the student of a particular point of view in the philosophy of science; hence the dialectical style of the work. We try to move argumentatively from one position to another without "killing off" the position from which we move.

Explanation is divided into three parts. In the first (Chapter 2) we ask the student to distinguish between science and pseudoscience as a way of figuring out what minimally qualifies something as a scientific explanation in the first place. Of course, what seems to be an easy task turns out to be a hard one, and without resolving it we move on to the second part (Chapters 3 through 8), trying to decide what makes one scientific explanation *better* or more adequate than another. Beginning with the view that all adequate explanations are causal explanations, we explore the limitations of that Aristotelian and Cartesian view by exploring Hume's criticism of causality and of induction (Chapter 3). After suggesting Kantian and rationalistic rejoinders to Hume's attack, we suggest another plausible follow-up to his work: explanation as description. That allows us to introduce the positivistic tradition and, in turn, to criticize it (Chapter 4). The covering law model (Hempel) is seen as a possible way to avoid the causalist/positivist stalemate, and its criticism allows the introduction of Michael Scriven's work and, through him, Wittgenstein's (Chapter 5).

By this point in the text, it is clear to most students that philosophical positions have relevance to science. We therefore move to some classical theories of epistemology and metaphysics in Chapter 6, and treat such subjects as the reality of universals, materialistic reductionism, and the mind-body problem. In Chapter 7 we provide a pragmatist "alternative" to traditional metaphysical epistemological language; and we finish this second part of the text with a summary of three broadly different contemporary views of the nature of explanation: the realist, empiricist, and relativist views (Chapter 8).

The third part of the text asks students to consider the relation of scientific explanation to other forms of explanation—in history (Chapter 9), in ethics (Chapter 10), and in religion (Chapter 11). Some of the literature is fruitfully parallel (for example, the Carl Hempel/Michael Scriven debate in the philosophy of history parallels that in the philosophy of science). The topic also allows students to see whether their views amount to saying that all explanation worth its salt must be scientific, or to figure out the ground rules for legitimizing different kinds of explanation without allowing just everything. Human freedom and determinism and the relation of religion to science are central themes in this part of the text.

Idiosyncrasies and Pedagogy

Why organize the entire text around the concept of explanation? Is too much of traditional concern to philosophers of science left out? We have found the concept so central that in some way or other most typical philosophy of science topics can be dealt with, and we think helpful entry points to some important nontraditional topics are provided. For example, the relationship between science and religion is of great importance both historically and for many contemporary students (including many who don't initially want to talk about such matters); yet the topic finds no place in most texts in the philosophy of science.

Our guiding principle for the selection of readings and for organization has been the desire to engage students actively in the debate and then keep them engaged by natural transitions to further positions. To that end we have sacrificed some of the look and balance and historical sequence that other texts typically exhibit. For example, we've not tried to give philosophers space in keeping with their greatness; instead the selections are measured by their ability to exhibit a position clearly and yet concisely enough for it to be appealing. While there is a rough chronological order to the positions as presented, colleagues will notice and may want to point out the anomalies. In some cases we have quoted a philosopher at length (e.g., Hume, in Chapter 3) and then followed with only our summary of an alternate position of great importance (e.g., Kant, following Hume). Clearly, no sense of philosophical fair play is used; we simply have found these different modes or presentations best in making a position understandable and believable.

Pedagogy also dictated the use of Velikovsky as a test case in Chapter 2, though the creation/evolution debate might seem more natural, more familiar to students, and more current. We have found that students benefit from an introduction to the issue of pseudoscience that is intriguing but of lower emotional temperature. Once students have developed some of their own ideas concerning scientific explanation, they can use their interest in the issues of science and religion in Chapter 11 more effectively.

Finally, in Chapter 8, empiricism, realism, and relativism are suggested as three contemporary views of scientific explanation. Their treatments are unequal, and this is not justified by their strengths or their contemporary prominence. Rather, we have suggested criteria that complete theories should meet, then worked out the first one as a sample, and hope that the reader will contribute more to each successive candidate theory. For helpful samples of the last two theories, we have referred students to Polanyi's *Science, Faith, and Society* and to Kuhn's *The Structure of Scientific Revolutions*, respectively.

Our purpose in this text is to open students to the rich debate now going on among philosophers of science. We want students to share in the fun and also to recognize the serious consequences. We want them to see how Kuhn and Mach could have gotten the same reading from a spectrophotometer and thus how on one level philosophy and philosophers could be ignored. But we also want students to know how fundamentally far apart a Bohm and a Heisenberg can be, as well as a Lysenko and a Vavilov. There are more traditional approaches than this, and they have their merits. We have tried to angle our approach and organization to the way students respond, get excited and engaged, and learn. We hope our colleagues find that this text serves their students in these ways.

Thanks

We are grateful to the many Earlham students who have used and criticized earlier versions of chapters and sections in this text. They have inspired, challenged, threatened, cajoled, and helped us to understand the philosophy of science better. We hope we have done the same for them. We also want to acknowledge the support of Earlham College through the Professional Development Fund. At crucial times Lois McMahan has provided the expertise and patience necessary for the preparation of the manuscript. Finally, Barbara and Mary Jo have believed in us and this book in the face of convincing evidence to the contrary. Thank you all.

Gerald Bakker
Len Clark

/1/ WHAT IS EXPLANATION?

EXPLAINING IS A COMMON but also very complex activity. We all engage in it daily and in a variety of ways. A two-year-old surrounded by spilled milk may say, "Teddy bear did it," and thus use explanation to divert parental displeasure. A chemist may use a half-million-dollar mass spectrometer to determine molecular structure, and then explain an observed reaction. One diverse and fascinating type of explanation is *scientific* explanation, which is the topic of this book. We will begin our exploration of this topic with a homely example.

The Case of Martin E.

Martin E., apparently happy and healthy husband, father of two teenagers, and pharmacist in a small midwestern city, has recently celebrated his fortieth birthday. To the dismay of his family and friends, Martin begins to exhibit increasingly disturbing behavior. Sleep-shattering nightmares are followed by increasing nervousness. He seems to be either ecstatically happy or depressed and in torment. Mildly self-destructive actions give way to a suicide attempt. At this point, Martin is hospitalized.

Why did all this happen? As medical and psychiatric experts, Martin's wife and children, his friends and business associates all begin to seek explanations, where will they look? And what principles, scientific or otherwise, will aid them in their search?

Let's consider the bewildering array of possibilities. They might include the following:

1. Martin has been ingesting harmful drugs readily available to a pharmacist.
2. Martin is having a mid-life crisis—a crisis of confidence in his professional achievement, his personal worth, all made more severe by increasing awareness of his own mortality.

1

3. Martin has absorbed harmful drugs through his skin, after handling prescription medicine in the course of his professional duties.

4. Some childhood trauma, its effects triggered by a current experience, has led to psychotic manifestations.

5. Martin has had organic, perhaps brain, damage due to excessive radiation—perhaps from nuclear tests—to which he is particularly susceptible.

6. Martin has an inherited mental disease.

7. Inadequate nutrition—perhaps a vitamin deficiency—has led to a malfunction of Martin's nervous system.

8. Martin has been driven mad by the carping of his wife and in-laws.

9. Martin is possessed by the Devil.

"Well," we might say, "now that we have a list of possibilities, let's amass some data, do a few experiments, and find out what the explanation is." It is wonderful that in many cases when explanations are sought we can do just that. Is the light bulb burned out, or is a fuse blown? We try a fresh bulb and if it works, our question is answered. Many explanations in science work that way, too.

But it is apparent that these straightforward procedures won't do in the case of Martin E. For one thing, some of the suggested explanations may overlap in complicated ways. Will the presence of certain unusual chemicals in the blood exclude the childhood trauma explanation? Will evidence of other organic radiation damage mean that radiation, and not a mid-life crisis, led to the suicide attempt? That is, one has trouble excluding certain explanations on the basis of experiment and observation. It seems, in fact, as if assumptions that make one or another explanation plausible may insulate them against counterevidence. What would count as convincing evidence that no devil was involved if you believe that devils do often involve themselves in people's affairs and artfully conceal their presence?

The problems don't stop here. Assumptions do influence what explanations are offered and what evidence is accepted as relevant. But so do interests and plans. For Martin and his family, for example, all explanations that do not lead to a return to health will be suspect. If one expert suggests that radiation is the culprit and that there is nothing to be done, while someone else suggests a reversible vitamin deficiency, Martin is likely to act as if the vitamin explanation is correct. On the other hand, the pharmaceutical company lawyers will surely be guided by what they hope to accomplish: a limitation of their client's liability. If evidence points to drugs absorbed by skin contact, they may well seek evidence for voluntary ingestion, or psychosis, etc., in an attempt to show that the drug absorption was neither a necessary nor a sufficient condition for the illness.

Finally, some philosophers and psychologists may want to stress Martin's freedom of choice; after all, as a free and responsible human being, he voluntarily attempted suicide. Others, of course, will find in such a defense of freedom merely avoidance of deeper genetic and environmental compulsions.

Confronted by such an array of possibilities, we are often tempted to become either skeptical of our ability to know or cynical about what difference such disputes make, anyway. The skeptic in us will urge, "It all depends on your assumptions, and we can't prove them one way or another." Our cynical side may point out that the difference of opinion about explanations will go on long beyond the recovery or death of Martin E. and that decisions that affect his life may not be based on more than the prejudgment of physicians and the method of jury selection when the pharmaceutical company comes to trial.

But there is another side to our reaction. We do believe, in practice, that some explanations are better than others and even that some are correct and others misguided, irrelevant, or wrong. We also feel that we learn adequate explanations that we didn't know before. All this gives us some reason for optimism: perhaps we can learn to choose the appropriate explanations and explanatory frameworks in hard cases, such as the case of Martin E., by reflecting on how we judge, accept, and reject explanations in cases where we are clear. However we find tools to deal with the question, we are drawn to the importance of the question.

The study of what it means to explain Martin's behavior will involve asking a number of more specific questions such as these:

1. Do different explanations conflict with one another, or might several of them all be relevant?

2. Are all the explanations tested in the same way, or do we have to devise different criteria for judging different types of proposed explanations?

3. Can we be sure when we have the right explanation? How do we judge the odds that we are "reasonably" sure?

4. What assumptions do we have to make about human nature, the nature of knowledge, and the relations between theories and evidence in order to accept or rule out any of the explanations on our list?

Explanation and Science

This book deals with the nature of scientific explanation. We clearly need some *general theory* of adequate explanation to give us the principles for solving the case of Martin E. More broadly, in fact, we need such guidance for all the many cases in which we are offered more than one explanation of the same event.

What is explanation? We will investigate several proposed definitions and the general theories behind them, but all of them have this much in common: explanation is central to knowing just about anything. For example, one popular characterization of explanation holds that it is a way to learn about what is unfamiliar to us by relating it to what is already familiar. Understood this way, seeking explanations is an important way to learn something new. Not only is explanation related closely to knowing; perhaps even more important, it is an access to *worthwhile* knowing. We seek explanations because of our need to know, because an event holds promise or danger, and we want to gain that control over it which explanatory knowledge may provide. We need to distinguish between this fertile kind of knowing, then, and other types of knowing—for example, classifying, or cataloguing, or experiencing—which may be interesting and important to us or may not. We seek to explain, and therefore the tools for choosing adequate explanations are powerful tools for knowing.

Our specific guiding question in this book will be *What constitutes an adequate scientific explanation?* Just as explanation in general is an important part of knowing, so scientific explanation is an activity central to the work of scientists. Relating explanation to some other scientific activities will illustrate its importance.

1. *Laws.* Scientists seek regularities, especially those invariable and most universal regularities that are called laws of nature. These laws are valuable when we use them to explain events as instances of universal relationships. The usefulness of particular kinds of laws will be judged, therefore, in part because of their explanatory power; and those judgments will depend on what we recognize as adequate explanations in general.

2. *Evidence.* Before seeking evidence, either by deducing facts from a theory or by experiment or observation, a scientist must have a clear idea about what the relevant evidence will be like. But the criteria for relevance of evidence will depend on the type of law or phenomenon being investigated, and that in turn will depend on what explanatory power the law or phenomenon promises. For example, if weather conditions are held to be an explanation of certain interesting behaviors in beavers, careful records of barometric pressure readings may become worthwhile biological data.

We also make important assumptions about the transferability of explanations when we assess the value of evidence. Research on carcinogens in animals is based partly on our belief that explanations of animal cancer will parallel explanations of human cancer.

3. *Theories.* Theories are valued largely for their explanatory power and simplicity, though we shall see that they are valued for other features as well. The choice of some theories over others often depends on what we want them to do—whether to predict, or to relate disparate research results, or to suggest probable sources for laws (all of these are related, as we will see). All of these criteria come from our search for explanations and therefore from our judgments about what sorts of explanations are best. If to explain most adequately is to predict with minimum complexity, for example, Copernican astronomy is unquestionably a better explanation of planetary movement than is Ptolemaic astronomy. If, however, to explain best in astronomy is to give an account based on the simplest geometric figures, no matter how complex the resulting calculations for prediction, or if navigational convenience is the criterion, Ptolemaic astronomy has the edge. We are often unaware of the complex web of explanation criteria we are using because they may be routinely assumed in our particular area of research. We become most aware of them when things aren't going well, when we are confused by our research results, or when a current theory, while understandable, is no longer helping us learn what we most want to know.

Newton's theory of gravitation left unexplained how one body could influence other bodies across stretches of empty space. For many thinkers at the time, this made the theory less adequate than the Cartesian explanation of motion, which denied the existence of empty space. But Newtonianism triumphed because of its power of mathematically predictive explanation. In this way, judgments about what an adequate explanation ought to do decided the course of classical physics.[1]

Explanation is clearly a central concept for the practice of science. Those who want to be as productive as possible as scientists will, at the very least, need a firm sense of what counts as an adequate explanation. It is so much the better if they have a clear and defensible theory of what adequate explanations look like.

Science and Philosophy

How does scientific explanation relate to explanation in general? Are there important kinds of explanation, legitimate and useful in their own right (for instance, historical or religious or "commonsense" explanations) that are not scientific and are judged by different criteria than are scientific explanations? Or will any explanation that is really adequate be, by that very fact, a scientific one? This question clearly cannot be decided solely by scientific criteria themselves, for then one "side" of the possible conflict will be the judge. The character of explanation itself, and the comparison of its various subtypes, is of special concern to that branch of philosophy known as *epistemology*, or the study of knowing. Epistemologists have, through the centuries, attempted to clarify not only the meaning of explanation but also the meaning and relationships of concepts such as knowing, truth, validity, and the like. Some of these epistemologists have specialized in considering the problems of scientific knowing; and in the last few decades we have come to recognize this epistemological subfield as the philosophy of science.

With the above distinctions, we can no doubt recognize in practice the difference between philosophers who may deal with the meaning of explanation and specialists in the philosophy of science who, as specialists at least, will restrict their analyses to the goals and methods of science itself. But as we consider the question of explanatory adequacy in science, how do we know when we are doing science and when we are doing philosophy, and does it make a difference which we are doing?

Until about 200 years ago, there was no distinction between science and philosophy. What we now call science was then called natural philosophy and was contrasted only with moral philosophy, which meant the study of humans and human action. Since that time the word *science* has come to refer to the investigation of nature by use of experiment and observation. The philosophy of science concerns not the study of nature but the study of science itself, as part of the larger epistemological enterprise of studying knowledge itself in all its branches. As one might suspect, this science/philosophy division is far from clear-cut. After all, scientists themselves often consider the principles of their own fields and make judgments about the relationship among different fields of scientific inquiry; and philosophers cannot afford to ignore experimental results, the predictive power of some scientific theories compared to others, and so on. Often, one person will contribute to both fields; for example, in Chapter 4 we will consider the work of Ernst Mach, who was both a physicist and a philosopher of science.

We shall consider philosophy and science not as sharply different fields but as segments on a spectrum of types of inquiry. At one end of the spectrum we might put applied technology, which does not question or test scientific principles but assumes them in order to design practical applications. Adjacent to technology we might locate science, which deduces consequences from principles and reflects to some extent on its own methods. The philosophy of science, farther along the spectrum, concerns itself especially with questions of method and the logic of explanations and not much at all with deducing consequences for application. It will be part of the purpose of this book not to demarcate sharply between science and philosophy but to show how scientific concerns merge into the use of philosophical positions and arguments. We can therefore afford to carry arguments about the best sorts of explanations back to their assumptions without worrying about precisely when we cease doing science and start doing philosophy. It will be clear enough that our concerns are philosophical when we come to consider how scientific explanations relate to explanations offered in other fields.

The explicit topic of this book is scientific explanation, but its purpose goes beyond that subject. It is also intended as an introduction to philosophy. We will find that the concept of explanation is so central to the work of classical philosophers that it provides an entry point to philosophy as a whole.

Our Plan of Study

Scientists choose among competing scientific explanations of phenomena. However, some purported explanations don't get considered in these choices because they are thought not to meet even the minimum standards for qualifying as scientific. Yet they are *asserted* to be scientific by their defenders. These cases raise the problem of defining *pseudoscience*. How can scientists judge that some explanations can be safely ignored? What are the minimum requirements for an explanation to be worthy of scientific testing? In Chapter 2 we will take up these questions.

Once competing explanations are suitably qualified for scientific consideration, we still need criteria for choosing among them. The grounds for this choice have been a major subject of the great classical theories of scientific explanation, and we will consider several of these theories, and what they recommend, in chapters 3 through 8.

Finally, we will find that these theories of scientific explanation have intricate and important relations to theories of appropriate explanation in other fields: history, ethics, law, social science, and religion. In some cases, scientific method has been suggested as a model for inquiry in these other fields. In other cases, scientific explanation has appeared to compete with other explanatory frameworks, and the results have been major disputes that continue to be among the most important problems facing thoughtful people. We will consider these issues in chapters 9 through 11.

We hope that the reader will be an active participant in the conversation of this book. The style is intended to foster this engagement and accounts for several unusual features of the work.

1. The organization is dialectical in the sense that a new theory of explanation is offered when problems with a previous theory suggest that other options be explored. Although the order is not always historically based, new positions are considered as they in fact developed in the philosophy of science in response to the problems of other positions.

2. Even so, the movement from one theory to another is not necessarily from a false position to a truer one. For example, in Chapter 4 we consider a theory of explanation known as positivism. Problems with that position have led some theorists to modify or abandon it and to seek other theories of explanation, which are considered in subsequent chapters. But problems are not all of equal weight, and not all problems are fatal to a position. Therefore, we present arguments but do not conclude in this work that positivism or any other classical theory is false. This style is true to the fact that there are serious thinkers who hold versions of the positivist position today. For them, the problems of other positions outweigh the problems of their own.

3. We, the authors of this volume, have our own positions on the issues, and we do not always agree. We have tried to keep our personal positions to ourselves and let the reader assess the strengths and weaknesses of the various theories unfettered by our preferences. Nevertheless, the very organization of chapters and topics within chapters is the result of judgments, at least of the importance, if not the validity, of some arguments. The best insulation from such subtle slanting of the issues is to use this work only as a beginning and to read the original sources that are cited and quoted here.

4. Examining original sources will reveal that we have not always presented arguments in the order in which they were presented by their authors. It will also become clear that some major actors on the world philosophical stage receive in this book less attention than some bit players. This book is not to be read as a history of philosophy but as an attempt to open philosophical questions for science students asking about how science works.

5. We have assumed in writing this book that the reader has moved far enough into a particular branch of science to know that introductory science texts don't tell the whole story. Texts usually give the conclusions of scientific work and explain the concepts. What is left out is discussion of the twists and turns of how scientists arrived at their conclusions. And almost never is there any analysis of the current major debates in the field. We assume that the reader knows science is not the cleaned up version presented at the first year or two of college study. We also assume that the reader has not had much formal study in philosophy. Finally, we assume that the reader is seeking to establish his or her own views on the questions we raise, and we therefore try to be as persuasive as

we can be on many different sides of an argument. To our students and to the reader we say, "The position of importance, the last word in the argument is not ours but yours."

Supplementary Readings*

Nagel, Ernest. *The Structure of Science*. New York: Harcourt, Brace and World, 1961.

Though this work is old, it is a standard in its field and worth using as a reference. Its writing style and analysis of the issues are substantial.

Edwards, Paul, ed. *Encyclopedia of Philosophy*. 8 vols. New York: Collier Macmillan Publishers, 1967.

This is a major source that any serious student of philosophy will use frequently.

Harré, Rom. *Great Scientific Experiments*. Oxford, England: Oxford University Press, 1983.

In the introduction the author writes, "I vividly recall the night my father and I prepared bromine." His story catches something of the fundamental importance of the physical phenomena, suggesting that however complex our mathematical laws, however elegant our models, however abstract our philosophical analyses, it is the experiment that is important. Harré describes twenty scientific experiments.

Note

1. In this text we will use the word *theory* to indicate something more than a guess or speculation. We will *not* use the word in the sense of "I have a theory the butler did it." A theory for scientists is something like the kinetic theory or relativity theory. Theory implies something more general, something at a higher or broader level than law.

*The order of presentation of the supplementary readings reflects the authors' suggestions for where to begin.

/2/ DISTINGUISHING SCIENCE FROM PSEUDOSCIENCE

WE BEGAN OUR ANALYSIS of scientific explanation by looking at different ways to explain Martin E.'s strange behavior. But not all the explanations listed would be classified as scientific explanations; some would be called *pseudoscientific*. Their proponents might defend them, but others, and perhaps most scientists, would want to dismiss them as not even worthy of consideration. We would do well, then, to inquire at the outset about which attempts will qualify as scientific and which ones will not.

We do not have to look far to find examples of explanations that are commonly called pseudoscientific. Horoscopes in the daily papers, articles in the *National Enquirer*, and a number of diet books offer explanations that claim to be "scientific," but almost all scientists would agree that they are not. It will be more instructive to study a proposed explanation or theory about which there has been significant disagreement and public debate.

Velikovsky: Science or Pseudoscience?

In the past thirty years no theory receiving any significant support has been so vigorously denounced as pseudoscientific as have the theories of Immanuel Velikovsky. His ideas regarding the recent geological history of the planet Earth have stirred strong opposition from among scientists, and in fact the label pseudoscientific is one of the gentler epithets used to describe his work. We shall proceed with an analysis of the arguments for and against Velikovsky's ideas on the assumption that often the most strongly debated cases elicit the most interesting questions.

9

Walter Sullivan, *New York Times* science editor, has provided some background information on Immanuel Velikovsky that should help to put his ideas in
some perspective.[1]

Velikovsky, by any mode of measurement, is an extraordinary man. He was
born in Vitebsk, Russia, in 1895, and decided early in life to study medicine. This was difficult, for he was Jewish and in Czarist Russia medical
education for a Jew was hard to come by. He went briefly to France to
study and, after a visit to Palestine, continued his pre-medical education
in Edinburgh. On a visit to Russia in 1914 he was trapped by the outbreak
of World War I and resumed his medical training there, receiving his degree from the University of Moscow in 1921.

His heart, however, was almost as much with Jewish culture as with
medicine. While he was doing his postgraduate studies in Berlin, from
1921 to 1923, he met Chaim Weizmann, later to become Israel's first
president, who had undertaken the establishment of a Hebrew University
in Jerusalem. Velikovsky joined in this effort by co-editing two series of
volumes, the *Scripta Universitatis*, containing articles by Jewish scholars
and published on behalf of the burgeoning university in Jerusalem. The
mathematical-physical section of the series was edited by Albert Einstein.

Weizmann, it seems, asked Velikovsky to start setting up the university
in Jerusalem, but he turned down the proposal, apparently not relishing
the prospect of intensive fund-raising and administration that this would
entail. However, late in 1923 he and his wife moved to Palestine, where
he practiced medicine and began studying psychoanalysis. He met and
corresponded with Freud and contributed an article to *Imago* (the psy-
choanalytic journal that Freud published in Vienna) which was later published in English under the title "Tolstoy's Kreutzer Sonata and
Unconscious Homosexuality." Velikovsky even decided to analyze Freud
himself, so to speak, writing on "The Dreams That Freud Dreamed." He
also was the first—or at least one of the first—to recognize the importance
of an electroencephalogram (EEG), or printout of electrical impulses from
various parts of the brain, in diagnosing epilepsy. When he read Hans
Berger's pioneering paper on 1929 on the monitoring of electrical emis-
sions from the brain, Velikovsky saw its application to epileptic attacks,
whose "lightning-like" onset he compared to the effects of an electric
short circuit. In a paper prepared in 1930 and published the following year
he urged the study of epileptic seizures with an EEG and suggested the
possibility of diverting, from the brain, the rapid electrical fluctuations
that, he suspected, were involved.

A major change in the course of his career occurred when he and his
family moved to New York in 1939 to further his research on a book dealing with Freud's three heroes: Oedipus, Akhenaton, and Moses. As Velikovsky began delving into ancient Egyptian as well as Hebrew texts, some
of the biblical catastrophes described in *Exodus*—such as the rain of fire,
plague of darkness, and parting of the Red Sea—seemed also to be re-

flected in the Egyptian writings. Might these, he asked himself, have been worldwide events of some terrible sort? There followed a research undertaking of formidable dimensions. He examined ancient chronicles from pre-Columbian America, China, India, Iran, Babylon, Israel, Egypt, Iceland, Finland, Greece, and Rome.

In many of them he found accounts of catastrophes that he decided had occurred coincidentally throughout the world. Finally, to explain them, he devised the admittedly extraordinary theory that the earth, during the period covered by these traditional accounts, had gone through a succession of cataclysmic encounters with comets and planets. The chief villain was Venus, which, he concluded, had been thrown off by Jupiter in the form of a comet that then flew an eccentric orbit, twice bringing it near the earth. It was an idea that had features in common with one proposed by Howard B. Baker, in a series of articles beginning in 1911. As noted earlier, Baker had suggested that the earth and Venus came close enough for the gravity of Venus to tear the moon from what became the Pacific Basin, setting the continents in motion to open up the Atlantic Basin. His hypothetical encounter with Venus came in the Miocene Period, some 20 million years ago, whereas Velikovsky saw it as much more recent, accounting for many of the Biblical catastrophes. Hydrocarbons in the form of naphtha from the "comet tail" of Venus, he said, fell on the earth, causing a rain of fire; the earth's spin axis tumbled so that the sun seemed to stand still in the sky, as recounted in *Joshua*; seas swept the lands and the Red Sea was drained briefly. Venus, at one point, collided with Mars, he said, which also repeatedly came near the earth before the planets settled into their present orbits.

Velikovsky interwove his interpretations of ancient history with recent scientific discoveries. The fact that the earth's magnetic polarity occasionally has reversed itself—a phenomenon that would cause the north-pointing needle of a compass to point south—he saw as evidence that when comets or planets almost collided with this planet, electric discharges took place between the two bodies, producing lightning bursts sufficient to reverse the earth's magnetic polarity.

Anyone interested in reaching a careful judgment about the merits of Velikovsky's reconstruction of events from the recent historical past must read his books *Worlds in Collision* and *Earth in Upheaval*.[2] However, an excellent summary of the events proposed by Velikovsky has appeared in the magazine *Pensée*. In reading this account it should be kept in mind that Velikovsky's primary purpose was to write a record of historical events using the writings of ancient peoples. His initial challenge was not to science and scientific law but to the interpretation given the writings and records left by ancient peoples. This, however, led him to challenge much of what has been said by physicists, astronomers, geologists, and biologists, and Velikovsky did not back away from the confrontation. He demanded a reconstruction of scientific laws as well. The *Pensée* summary follows a short editorial introduction:[3]

Immanuel Velikovsky manifests a strong distaste for summaries and popularizations of his books. In the past, many erroneous criticisms of his work have been based upon such popularizations, the critics never having studied his books. And indeed these books, detailed in their argument and exhaustive in their documentation, do not easily lend themselves to summarization.

Nevertheless, in embarking on a project designed to give the fullest possible coverage of all aspects of Velikovsky's work, the editors of Pensée felt it desirable to reacquaint readers with the flow of events described in his revolutionary reconstruction of recent solar-system history. The evidence, amassed in Worlds in Collision *and* Earth in Upheaval, *is not presented here, and to those who have not read these works the events must necessarily appear fanciful and insupportable. This difficulty can be remedied, of course, only by direct reference to the scholarly, evidential texts of Velikovsky himself.*

The following brief sketch was prepared entirely without Velikovsky's help. The serious student and scholar should resort to a careful reading of Worlds in Collision *and* Earth in Upheaval.

Global cataclysms fundamentally altered the face of our planet more than once in historical times. The terrestrial axis shifted. Earth fled from its established orbit. The magnetic poles reversed.

In great convulsions, the seas emptied onto continents, the planet's crust folded, and volcanos erupted into mountain chains. Lava flows up to a mile thick spilled out over vast areas of the Earth's surface. Climates changed suddenly, ice settling over lush vegetation, while green meadows and forests were transformed into deserts.

In a few awful moments, civilizations collapsed. Species were exterminated in continental sweeps of mud, rock, and sea. Tidal waves crushed even the largest beasts, tossing their bones into tangled heaps in valleys and rock fissures, preserved beneath mountains of sediment. The mammoths of Siberia were instantly frozen and buried.

Surviving generations recorded these events by every means available: in myths and legends, in temples and monuments to the planetary gods, precise charts of the heavens, sacrificial rites, astrological canons, detailed records of planetary movements, and tragic lamentations amid fallen cities and destroyed institutions.

"ALL IS RUIN"

Aware of a link between the circuit of heavenly bodies and the catastrophic ruin of previous generations, the ancients ceaselessly watched the planetary movements. Their traditions recalled that when old epochs dissolved, the new "Age" or "Sun" was marked by different celestial paths. Astronomers and seers diligently watched for any change which might augur approaching destruction and the end of an age.

Prior to the second millenium B.C. ancient Hindu records spoke of four visible planets, excluding Venus. Babylonians, meticulous in their observations, likewise failed to notice Venus.

But some time before 1500 B.C., Jupiter, for centuries chief among the deities, shattered the serenity of the skies. A brilliant, fiery object, expelled from that planet, entered upon a long, elliptical orbit around the sun. The feared god Jupiter had given birth to the comet and protoplanet Venus.

Terrified, men watched the "bright torch of heaven" as it traversed its elongated orbit, menacing the Earth. Venus, a Chinese astronomical text recalls, spanned the heavens, rivaling the sun in brightness. "The brilliant light of Venus," records an ancient rabbinical source, "blazes from one end of the cosmos to the other."

The fears of the star-watchers were justified. As Venus arched away from its perihelion during the middle of the second millenium B.C. *(ca.–1450), the Earth approached this intruder, entering first the outer reaches of its cometary trail. A rusty ferruginous dust filtered down upon the globe, imparting a bloody hue to land and sea. The fine pigment chafed human skin, and men were overcome by sickness. Those who sought to drink could not. Rivers stank from the rotting carcasses of fish, and men dug desperately for water uncontaminated by the alien dust. "Plague is throughout the land. Blood is everywhere," bewailed the Egyptian, Ipuwer. "Men shrink from tasting, human beings thirst after water. . . . That is our water! That is our happiness! What shall we do in respect thereof? All is ruin."*

As recalled by the Babylonians, the blood of the celestial monster Tiamat poured out over the world.

But as the Earth's path carried it ever more deeply into the comet's tail, the rain of particles grew steadily more coarse and perilous. Soon a great hail of gravel pelted the Earth. " . . . there was hail, and fire mingled with hail, very grievous, such as there was none like it in all the land of Egypt since it became a nation." So recorded the author of Exodus.

Fleeing from the torrent of meteorites, men abandoned their livestock to the holocaust. Fields of grain which fed great cities perished. Cried Ipuwer, "No fruits, no herbs are found. That has perished which yesterday was seen. The land is left to its weariness like the cutting of flax." These things happened, say the Mexican Annuals of Cuauhtitlan, when the sky "rained, not water, but fire and red-hot stones."

As our planet plunged still deeper into the comet's tail, hydrocarbon gases enveloped the Earth, exploding in bursts of fire in the sky. Unignited trains of petroleum poured onto the planet, sinking into the surface and floating on the seas. From Siberia to the Caucasus to the Arabian desert, great spills of naphtha burned for years, their billows of smoke lending a dark shroud for human despair. Our planet was pursuing a near collision course with the massive comet's head.

Suddenly, caught in an invisible grip, the Earth rocked violently; its axis tilted. In a single convulsed moment, cities were laid waste, great buildings of stone leveled, and populations decimated.

"The towns are destroyed. Upper Egypt has become waste. . . . All is ruin. . . . The residence is overturned in a minute." Around the world, oceans rushed over mountains and poured into continental basins. Rivers flowed upward. Islands sank into the sea. Displaced strata crashed together, while the shifting Earth generated a global hurricane which destroyed forests and swept away the dwellings of men.

In China waters "overtopped the great heights, threatening the heavens with their floods." Decades of labor were required to drain the valleys of the mainland. Arabia was transformed into a desert by the same paroxysms which may have dropped the legendary Atlantis beneath the ocean west of Gibraltar.

With dulled senses, survivors lay in a trance for days, choking in the smoky air.

The tilting axis left a portion of the world in protracted darkness, another in extended day. From the Americas to Europe to the Middle East records tell of darkness persisting for several days. On the edge of the darkness, the peoples of Iran witnessed a threefold night and a threefold day. Chinese sources speak of a holocaust during which the sun did not set for many days and the land was aflame. Peoples and nations everywhere, uprooted by disaster, wandered from their homelands.

CELESTIAL DRAGON

Led by Moses, the Israelites fled the devastation which brought Egypt's Middle Kingdom to an end. As they rushed toward the Sea of Passage, the glistening comet, in form like a dragon's head, shone through the tempest of dust and smoke. The night sky glowed brightly as the comet's head and its writhing, serpentine tail exchanged gigantic electrical bolts.

The great battle between the fiery comet's head and the column of smoke—between a light-god and a leviathan serpent—was memorialized in primary myths around the Earth. Babylonians tell of Marduk striking the dragon Tiamat with bolts of fire. The Egyptians saw Isis and Set in deadly combat. The Hindus describe Vishnu battling the "crooked serpent." Zeus, in the account of Appollodorus, struggled with the coiled viper Typhon.

The fugitive Israelites, having reached Pi-ha-khiroth, at the edge of the Red Sea, were pursued by the Pharoah Taoui-Thom (Typhon). The great sea lay divided before the slave people, its waters lifted by the movement of the Earth and the pull of the comet. Crossing the dry sea bottom, the Israelites escaped from Egypt.

As the comet made its closest approach to Earth, Taoui-Thom moved his armies into the sea bed. But even before the entire band of Israelites had crossed to the far side, a giant electrical bolt flew between the two planets. Instantly the waters collapsed. The Pharoah, his soldiers and chariots, and those Israelites who still remained between the divided waters were cast furiously into the air and consumed in a seething whirlpool.

The battle in the sky raged for weeks. A column of smoke by day, a pillar of fire by night, Venus meted destruction to nations large and small. To the Israelites, it was an instrument of national salvation.

Through a series of close approaches, the comet's tail, a dreadful shadow of death, cinctured the Earth, wreathing the planet in a thick, gloomy haze that lasted for many years. And so, in darkness, a historical age ended.

Possibly the human race would have become extinct, but for a mysterious, life-giving substance precipitated in the heavy atmosphere—the nourishing "manna" and "ambrosia" described in the ancient records of all peoples. It fell with the morning dew, a sweet, yellowish hoar frost. It was edible. The ambrosial carbohydrates, possibly derived from Venus' hydrocarbons through bacterial action, filled the atmosphere with a sweet fragrance. Streams flowed with "milk and honey." When heated, this "bread of heaven" dissolved, but when cooled, it precipitated into grains which could be preserved for long periods or ground between stones. Its presence allowed man and beast to survive.

In the new age the Sun rose in the east, where formerly it set. The quarters of the world were displaced. Seasons no longer came in their proper times. "The winter is come as summer, the months are reversed, and the hours are disordered," reads an

Egyptian Papyrus. The Chinese Emperor Yahou sent scholars throughout the land to locate north, east, west, and south and draw up a new calendar. Numerous records tell of the earth "turning over." An Egyptian inscription from before the tumult says that the Sun "riseth in the west."

While men attempted to determine the times and seasons, Venus continued on its threatening course around the Sun. Under Joshua, the Israelites had entered the Promised Land, and again Venus drew near. It was while the Canaanites fled from before the hand of Joshua in the valley of Beth-horon—some fifty years after the Exodus—that the daughter of Jupiter unleashed her fury a second time. "The Lord cast down great stones from heaven upon them unto Azekah, and they died." The terrestrial axis tilted. Once more the Earth quaked fiercely. Cities burned and fell to the ground. Above Beth-horon the Sun stood still for hours. On the other side of the Earth, chroniclers recorded a prolonged night, lit only by the burning landscape. This occurred, Mexican records report, about fifty years after an earlier destruction. As in its first encounter with the young comet, the Earth's surface was torn with great rifts and clefts, and hurricanes scoured the land. Strata pressed against strata, rising thunderously into mountains or engulfing cities. But the Earth and some of its inhabitants survived.

Anticipating renewed devastation following another fifty-year period, nations bowed down before the great fire goddess. With bloody orgies and incantations they enjoined the dreaded queen of the planets to remain far removed from the human abode. "How long wilt though tarry, O Lady of heaven and earth?" inquired the Babylonians. We sacrifice unto Tistrya," declared a priest in Iran, "the bright and glorious star, whose rising is watched by the chiefs of deep understanding."

MARS

In both hemispheres men fixed their gaze anxiously on the comet as, for centuries, it continued its circuit, crossing the orbits of both Earth and Mars. Before the middle of the eighth century B.C., astrologers observed dramatic irregularities in its wanderings. Viewed from Babylonia, Venus rose, disappeared in the west for over nine months, then reappeared in the east. Dipping below the eastern horizon, it was not seen again for over two months, until it shone in the west. The following year Venus vanished in the west for eleven days before reappearing in the east.

But this time it was Mars, not Earth, that endured a cosmic jolt. Passing by the smaller orb, Venus pulled Mars off its orbit, sending it on a path that endangered the Earth. A new agent of destruction was born in the unstable solar system.

This occurred in the days of Uzziah, King of Jerusalem. (Lucian, the Bamboo Books of China, the Hindu Surya-Siddhanta, the Aztec Huitzilopochtli epic, the Indo-Iranian Bundahish, etc., describe the reordering of Mars' and Venus' orbits.) Aware of the baleful meaning of irregular celestial motions, the prophet Amos, echoed by other observers of the sky, warned of new cosmic upheavals. Events soon vindicated the pessimistic seers.

As Mars drew near, the Earth reeled on its hinges. West of Jerusalem, half a mountain split off and fell eastward; flaming seraphim leaped skyward. Men were tossed into streets filled with debris and mutilated bodies. Buildings crumbled and the Earth opened up.

These cataclysms were associated with the founding of Rome (placed by Fabius Pictor at 747 B.C.) and with the death of Rome's legendary founder, Romulus. "Both the poles shook," Ovid relates, "and Atlas shifted the burden of the sky. . . . The sun vanished and rising clouds obscured the heaven . . ." Mars, the lord of war, became the national god of Rome.

Much smaller than Earth, Mars could not equal Venus in destructive power. But again the Earth altered its course around the Sun. The old calendar, with 360-day years and 30-day months, became outdated. Emperors and kings directed their astrologers to develop a new calendar.

BATTLE OF THE GODS

Mars and Venus now competed for the allegiance of men. Tribes moved from their homeland, confronting new enemies while petitioning Mars or Venus for a swift victory. Cities and temples were dedicated to the two planetary gods who determined the fate of nations.

The era of conflict between Mars and Earth and between Mars and Venus continued until 687 (or possibly 686) B.C. Hebrew prophets after 747 B.C. cried apocalyptically of upheavals yet to come. Reminding the Israelites of their passage out of Egypt, they declared that once more the whole Earth would quake, the moon turn to blood, the sun darken and the Earth be consumed in blood, fire, and pillars of smoke.

The catastrophe, as Mars hurtled past the Earth, came in the year 721 B.C., on the day Jerusalem's King Ahaz was buried. Under the influence of Mars' passage, the Earth's axis tilted and the poles shifted. Earth's orbit swung wider, lengthening the year.

Israelites observed the Sun hastening by several hours to a premature setting. Thereafter, the solar disc made its way across the sky ten degrees farther to the south.

Seneca records that on the Argive plain in Greece the early sunset came amid great upheaval. The tyrant Thyestes beckoned the entire universe to dissolve. The Great Bear dipped below the horizon. In the days which followed, states Seneca, "The Zodiac, which, making passage through the sacred stars, crosses the zones obliquely, guide and sign-bearer for the slow-moving years, falling itself, shall see the fallen constellations."

Once a peaceful, barely noticed planet, but now the "king of battle," Mars was still not finished with his work of destruction. In 687 B.C. a powerful Assyrian army led by Sennacherib marched toward Judah. On the evening of March 23, the first night of the Hebrew Passover, when Sennacherib and his army camped close to Jerusalem, Mars made a last, fateful approach to the earth. A great thunderbolt—a "blast from heaven"—charred the soldiers' bodies, leaving their garments intact. The dead numbered 185,000. Ashurbanipal, Sennacherib's grandson, later recalled "the perfect warrior" Mars, "the Lord of the storm, who brings defeat."

The same night of March 23, 687 B.C., in China, the Bamboo Books reveal that a disturbance of the planets caused them to go "out of their courses. In the night, stars fell like rain. The Earth shook." Romans would celebrate the occasion: "The most important role in the (Roman) cult of Mars appears to be played by the festival of Tubilustrium on the twenty-third day of March."

The Sun retreated by several hours. In certain longitudes the solar disc, which had just risen, returned below the horizon. In others, the setting Sun retraced its course, rising in the sky. The Hebrews witnessed the prolonged night of Sennacherib's destruction.

The Sun's retreat, due to a ten-degree tilt of the earth's axis, corrected the axis shift of 721 B.C. "So the sun returned ten degrees, by which degrees it was gone down," reads Isaiah 38:8.

From one continent to another men, oppressed with terror, watched Mars battle Venus in the sky, speed fiercely toward the Earth bringing blasts of fire, retreat and engage Venus once more. Perhaps the most startling literary account of this theomachy, or battle of gods, is contained in Homer's Iliad. (Velikovsky's revised chronology places Homer later than 747 B.C). As the Greeks besieged Troy, Athena (Venus) "would utter her loud cry. And over against her spouted Ares (Mars), dread as a dark whirlwind. . . . All the roots of many-founted Ida were shaken, and all her peaks." The river "rushed with surging flood" and "The fair streams seethed and boiled."

Mars was thrown out of the ring; Venus emerged a tame planet pursuing a near-circular orbit between Mercury and Earth. Where once it ranged high to the zenith, now it became the morning and evening star, never retreating more than forty-eight degrees from the Sun. Isaiah, who had witnessed the planet's destructive power, sang of its disgrace: "How art thou fallen from heaven, O Lucifer, son of morning! How art thou cut down to the ground, which didst weaken the nations! For thou hast said in thine heart, I will ascend into heaven, I will exalt my throne above the stars of God."

So ends the Pensée summary of Velikovsky's ideas. Though offered primarily as a historical account, this description of planetary events is also clearly intended to be a scientific explanation of how the planetary system developed. The response of the scientific community was swift and without equivocation. Velikovsky was denounced as a crank, and his work was treated as having no scientific value whatsoever. Macmillan Publishing Company, which had published Worlds in Collision as a scientific text, soon gave in to the pressure and transferred its publication rights to Doubleday. Although Harper's Magazine and Reader's Digest published articles favorable to Velikovsky's ideas, the reviews by scientists were universally damning.[4]

The defense of Velikovsky's work was also swift and vigorous. The treatment accorded Worlds in Collision was denounced as being unscientific, deliberately unfair, and the result of a fear of new ideas in a profession supposedly dedicated to looking for new ideas. Scientists were accused of engaging in a conspiracy to suppress Velikovsky's ideas. They were quoted as not having read Worlds in Collision and condemned for that fact and for admitting it. They were roundly criticized for not having given Velikovsky's ideas a fair hearing. One might suppose, after all, that any explanation purporting to be scientific deserves a complete, detailed, and careful analysis. Scientists could be expected to check the facts, examine the reasoning for logical errors, and see if the explanation is fruitful, that is, productive of correct predictions. It appears from the record, however, that very little of this kind of analysis was carried out on Velikovsky's work.

How did scientists decide that Velikovsky's work was pseudoscientific? Asking that question is asking for more than how Velikovsky's ideas were declared wrong. It is asking for the signals or criteria by which scientists decide that something is so wrong that it doesn't even deserve a full analysis. Every scientist is aware of the fact that many of the most striking advances in science are made not by the established authorities in the field but by people who are new to the area of research, either just beginning a scientific career or moving to a new field.[5] But not all new ideas are considered worth the time to analyze in detail. Velikovsky was not in fact treated as the originator of a striking advance but as a pseudoscientist whose ideas should be squelched or ignored. Scientists noted his credentials—he was not trained as a physicist or astronomer. They objected to the fact that Velikovsky went first to the public with his ideas (through popular magazines) and not to the scientific literature. They criticized his lack of under-standing of existing scientific thinking. The argument against Velikovsky was more than "He doesn't know what he is talking about." It was "He *couldn't* know what he is talking about."

Notice that both the detractors and supporters of Velikovsky are willing to accept the existence of reviewers and a review process, teachers and an educa-tional system, editors and a scientific literature. Velikovsky's supporters argue that the reviewers and teachers and editors are wrong but not that they ought not exist. These supporters are not asking that the general public be asked to decide for or against Velikovsky's ideas but for approval by the scientific commu-nity. Regardless of the outcome of arguments about quality of credentials and fairness, the first lesson to be learned from the examination of pseudoscience is that science is a functioning social system.

Thomas Kuhn has made an important contribution to our knowledge of science as a social system by looking at the history of science to see how in fact science has progressed.[6] We will come back to this "science as social system" theme later as we get further evidence that scientific explanation is more than just an exercise in logic. The role of the scientific community is a topic of considerable significance.

Let's briefly review the issue thus far. First, it is not clear that Velikovsky should be ignored just because he was not a physicist or astronomer. But it seems reasonable that reviewers' opinions and the judgment of science may count against a new idea.[7] Velikovsky and his adherents offered their explana-tion of planetary events for approval by the scientific community. On various grounds, some not related to the internal logic of the explanation, scientists have decided Velikovsky's ideas are pseudoscientific. One central question that emerges is the role of the scientific community in defining what is accepted as a scientific explanation.

What Counts as Evidence?

The first thing that strikes the casual reader of Velikovsky's works is the basis he used for his theories: ancient myths and legends as they have come down to us in the writings and other records that still exist. Velikovsky developed his theories of global catastrophe out of the legends of many ancient civilizations, and he argued that only actual global events can explain the literature that all around the world speaks of immense physical catastrophe. This central feature of Velikovsky's argument was attacked by scientists who argued that ancient myths and stories are notoriously unreliable and should not be taken as accurate records of geological and astronomical events. These writings primarily served religious, literary, and political functions, and so, according to the scientists, it would be sheer foolishness to read these as precise and accurate accounts of actual events. Even current newspaper accounts of such things as hurricanes and earthquakes contain exaggeration and hyperbole for dramatic effect. How could anyone possibly take seriously stories about gods and goddesses mixed up with natural events and national origins and try to read these as one does the U.S. Weather Bureau's report of yesterday's rainfall and wind velocity? Add to this the fact that many stories were passed down orally for years before being re-corded, and scientific reliability is even less likely. Therefore, concludes the argument, with much of Velikovsky's work crucially dependent on unreliable evidence, it would be pointless to take Velikovsky's theories seriously.

The response of Velikovsky to this argument came in three parts. First, he pointed to a remarkable parallelism in a number of accounts from widely sepa-rated peoples. Egyptian, Iranian, Chinese, and Mexican stories, just to name a few, tell of natural events that can be dated as occurring around the fifteenth and seventh centuries B.C. He granted that there are differences in the accounts but said the similarities are so striking as to be explainable only by global catastrophes that in the stories are clearly tied to astronomical events. We have been mistaken in assuming all along that ancient writings are devoid of scien-tific content. These writings refer to real events, and the job of the scientist historian is to find the common core of fact and try to understand what really happened.

Second, Velikovsky maintained that if scientists are going to be truly open to new ideas, they will, at times, have to be open to the possibility of working with new kinds of evidence. He believed this is one of those times—and a careful consideration of Velikovsky's ideas *will* require looking at all sorts of writings previously judged unreliable. If the scientists will set aside their preju-dices, he believed, they will find an ample basis for world catastrophes caused by the wanderings of Venus and Mars.

Finally, Velikovsky thought it is possible to explain the common view that ancient myths do not refer to real geological events, a view that predates our modern scientific culture, by positing a "collective amnesia." The events that resulted from the close interaction of Venus and Mars with the Earth were so terrible, so destructive to mankind that the only way ancient peoples could deal with the memories was to fictionalize the events and not have to relive in memory the terror and death and destruction.

At this point the debate began to warm up a bit. The three points just made are dealt with more sharply by Velikovsky's critics. Collective amnesia is called a nonexplanation; it is said not to be a general phenomenon, for there are no other instances of it; it is called a principle of ignorance, a purely ad hoc hypothesis, an untestable hypothesis. The second argument about openness to new types of evidence is seen by scientists to be only a variant of the first; so the crux of the matter becomes the core of actual events that can be obtained from the ancient writings. And, it is charged, Velikovsky did some totally unwarranted things with the texts to justify his interpretations. He had to be highly selective, he had to change accepted dates in Egyptian chronology, and he had to interpret words and phrases in his own way, all indicating that his own theories molded his interpretations of the texts. The ancient writings, though full of stories of holocausts and strange happenings, are far too ambiguous to support Velikovsky's theories about the interactions of Venus, Mars, and Earth. Here the thread of argument and counterargument becomes indistinct, with the supporters of Velikovsky speaking of unfairness and misrepresentations, and the scientists sighing and saying that they have seen enough to convince them further analysis would not be worth their time.

However, the supporters and opponents of Velikovsky seem to agree on two points, and these are relevant to our analysis of scientific explanation. The first point is that the relationship between observer, recorded observation, and phenomenon observed is a complex relationship but that the best data, or most useful scientific evidence, results when the observer is objective and the recorded observation is a faithful reproduction of what actually happened. Stated in these terms, finding this common ground between the two disputing camps seems hardly worthy of note. Yet we shall find out (beginning in Chapter 3) that this description of the best scientific evidence can be pushed to reveal a number of significant questions such as What is the relationship of an event and a perception of the event? What is entailed by moving from event to language describing the event, with only the latter being part of a scientific explanation? Is it really possible for an observer to be truly objective and not involved in the phenomenon observed? Think, for instance, of field studies in anthropology or of a physicist working with subatomic particles (such as the position and velocity of an electron). As these questions are dealt with in later chapters and new distinctions developed, the reader may find it profitable to refer back to the above argument on the use of ancient writings as evidence. We may find what could have been done to keep the two sides from talking past each other.

The second area of agreement that can be elicited from the argument on evidence is that scientific facts should be obtained and handled in a manner free of interpretive or theoretical bias. The scientists' charge that Velikovsky had manipulated the writings to favor his theories was responded to by "No, he didn't!" rather than "Of course he did. Why shouldn't he?" Setting aside any questions of motive or deliberate manipulation, we can see that facts were assumed by both sides to be ideal when free of theoretical bias. How true this is or in what sense it can be true are questions we shall find reasons to explore in Chapter 4 and later chapters.

One final comment on the question of what counts as evidence. It is not only between scientists and those they label pseudoscientists that questions arise about evidence. The opponents of Galileo did not want to admit as relevant what Galileo had seen of the moons of Jupiter through the telescope. The heavens, they argued, are immutable, and looking at Jupiter through this new instrument that distorts one's vision surely could not tell us anything about the celestial sphere. Another example can be found in the lung cancer/smoking controversy, where defenders of the tobacco industry argue that the statistical correlation of cancer incidence with cigarette smoking does not prove that smoking causes cancer. Arguments about evidence are not at all unusual.

Explanation and Prediction

The basis of Velikovsky's ideas in ancient writings remains a matter of dispute. Some find support for a series of global catastrophes, and others argue that Velikovsky's use of such evidence only indicates he did not know what good evidence is and therefore that none of his proposals merit attention. But no one disputes the fact that Velikovsky made a number of predictions that have subsequently turned out to be right. He predicted that Venus would be found to be hot, that Jupiter would emit radiowaves, and that the earth has a magnetosphere. These successful predictions, along with others,[8] are offered as evidence that Velikovsky's theories work and therefore should be accepted.

Because Venus has recently been active, argued Velikovsky, it should still be hot. Its recent formation from Jupiter and its brushes with Earth and Mars mean that Venus has not had time to cool down and should have retained some of its heat. When Velikovsky made this prediction, other cosmologists thought Venus would be cold and lifeless. The Pioneer space probe, however, has shown that Venus has a surface temperature of 460 degrees Celsius; Velikovsky was right and the others were wrong. If his theory leads to a correct prediction, then his theory must be correct—so the argument goes.

Prediction is important in science, and predictions that turn out to be successful often mark the turning point in acceptance of a theory that has been offered. For example, the chemical world was ready for a system of classification of the chemical elements at the time of Mendeleev, but his successful prediction of the existence and properties of three new elements is often cited as providing the final impetus toward acceptance of Mendeleev's Periodic Table. However, Velikovsky's success in his predictions has not changed the status of his theories with scientists. What is the matter with scientists? Are they so opposed to some new ideas that they will listen to no arguments in support of Velikovsky's theories?

Scientists have responded by saying that predictions that are not precise and detailed are not useful to anyone. They have argued that specific details are missing from Velikovsky's predictions. What was the source of heat? What has been the rate of cooling? What was the original temperature? And the current temperature? Without this kind of specific information it is impossible to test the prediction. A fortune-teller's prediction about a "tall, dark, and handsome stranger" is bound to come true unless some specifics of time and place and consequences are added. Fortune-tellers are successful precisely because they know how to be convincing without getting specific enough to be checked. So also with Velikovsky, say the critics; vague and imprecise predictions do not prove anything about Velikovsky's theories and explanations. In response, Velikovsky's defenders point to the number of correct predictions he made and describe him as a cosmologist of such broad vision that one could not expect him to be in the trenches doing the dirty work of checking out every detail.

What appeared earlier to be a simple case of "correct prediction, therefore correct theory" turns out to be much more complex. Suppose we have a theory that says all swans are white; this can be either a simple lawlike generalization or an elaborate genetic mechanism linking color integrally to the other swan characteristics. If we predict that the next swan we see will also be white, and in fact it is, how much have we added in proof that our theory is correct? Does one more confirming instance add very much? Of course not. We see another swan and it is white, and no one is much impressed. With a trivial example like this, it is clear that merely making a correct prediction does not confirm that one's theory is correct. But suppose one argues that Velikovsky predicted something that was unexpected and not predicted by those using the standard theories. Doesn't that strengthen Velikovsky's position? In one sense, it does not strengthen it. The logical force of an argument cannot be enhanced by surprise. Unexpectedness can affect the psychological impact of an argument but not its logical status.

Let's try another tack with the argument that successful prediction means a correct theory. Perhaps using phenomenon and explanation more like real science than white swan generalizations will bring us closer to a cosmological theory such as Velikovsky's. Let's take as a phenomenon to be explained, pellagra, a disease that was all too common in the southern United States in the early 1900s. Pellagra is characterized by rough, red skin eruptions that are followed by diarrhea, lassitude, dizziness, and mental disorders. One explanation offered for the cause of pellagra was that it was a hereditary disease. It was observed only among poor people, and the explanation accepted for years was that pellagra, low IQ, lack of motivation, and economic poverty were all evidence of inferior breeding stock.

How do we decide whether pellagra is inherited, and what role will prediction play in our decision? Testing or confirming a hypothesis is usually done in four ways: (1) increasing the amount of evidence, (2) obtaining a variety of evidence, (3) increasing the precision of the evidence, and (4) developing indirect theoretical support. In all four ways, the hypothesis that pellagra has a genetic basis can be given strong support. There is first the fact that pellagra and poverty could be observed together in many places and then the observation that no rich person ever came down with pellagra. Then it was observed that pellagra and poverty were associated in other countries and other times. It was also possible to link the incidence of pellagra with the extent of economic deprivation. Finally, it was shown that the children of pellagra victims were likely to develop pellagra, just as the children of low-IQ parents tended to have low IQs, and there are also other diseases such as sickle-cell anemia that have a genetic basis. A very strong case linking pellagra and a genetic defect was made in this way.

Successful predictions were made, and the hypothesis tying pellagra to a genetic defect was very strongly confirmed. The only thing wrong with this is that a few simple experiments can show the hypothesis to be false. Providing a well-balanced diet to pellagra victims can cure all the pellagra symptoms. It is now accepted that pellagra is a vitamin-deficiency disease. The theory that pellagra was a hereditary disease was well confirmed, but it was proved to be wrong. Successful predictions turned out not to be decisive.

Pseudoscience and Explanatory Criteria

We have seen that the question of whether Velikovsky was a pseudoscientist is more complex than it appeared at first. Though we have been able to isolate at least two of the major issues, namely, the definition of what is to count as evidence and the role of correct predictions, we did not find the argument settled. One important reason for this is that we do not yet have a full picture of how all these aspects of a scientific explanation's value *fit together*. Were we to explore all the dimensions of the debate about Velikovsky, we would have not two issues—evidence and prediction—in front of us, but many issues, each with those who would claim it as decisive.

In succeeding chapters, we will be reviewing the several issues that scientists and philosophers have seen as crucial in assessing an explanation as adequate or inadequate. In the process, the reader will, we hope, find criteria sufficient to return to the pseudoscience issue with tools at hand for making a decision. Certainly the fact that we have not found an answer that is clear and obvious, apart from the more detailed work that follows in this book, does not mean that most scientists are undecided. They believe Velikovsky was writing fiction, not science. The reader should be able to assess this judgment with some confidence when we have surveyed the various theories of the nature of scientific explanation on which it might be based.

We have in this chapter opened up questions about the relevance of the scientific community, the nature of evidence, and the relation of prediction to explanation. In Chapter 3 we shall explore these questions further by looking at the concept of causality, a fundamental concept in science. In the scientific community there is disagreement on the nature of causality and its role in scientific explanation. Just as with the fight between science and pseudoscience, we will try to develop differing positions but not settle the argument. That we leave to the reader.

Supplementary Readings

Goldsmith, Donald, ed. *Scientists Confront Velikovsky.* Ithaca, N.Y.: Cornell University Press, 1977.

This book includes some of the papers presented at a symposium on Velikovsky's views at the 1974 meeting of the American Association for the Advancement of Science.

Thagard, Paul. "Why Astrology Is a Pseudoscience." In *Introductory Readings in the Philosophy of Science,* edited by E. D. Klemke, Robert Hollinger, and A. David Kline. Buffalo, N.Y.: Prometheus Books, 1980, pp. 66–75.

This is a thoughtful look at the problem of how one distinguishes science from pseudoscience. Note how much importance Thagard places on the role of the scientific community.

Kitcher, Philip. *Abusing Science: The Case Against Creationism.* Cambridge, Mass.: MIT Press, 1982.

Kitcher has written well about the creation-evolution debate and how one can recognize pseudoscience.

Haven, Marsha P., Osler, Margaret J., and Weyant, Robert G., eds. *Science, Pseudo-Science and Society.* Waterloo, Ontario: Wilfred Laurier University Press, 1980.

This work is comprised of papers given at a symposium held at the University of Calgary. The treatment is substantial and covers the range of opinions from those who believe that the demarcation between science and pseudoscience is clear to those who maintain that distinguishing the two is no longer possible.

Gardner, Martin. *Fads and Fallacies.* 2d ed. New York: Dover Publications, 1959.

An easy-to-read discussion of pseudoscience.

Notes

1. Walter Sullivan, *Continents in Motion* (New York: McGraw-Hill, 1974), pp. 26–28.

2. Immanuel Velikovsky, *Worlds in Collision* (New York: Doubleday, 1950); and *Earth in Upheaval* (New York: Doubleday, 1955).

3. *Velikovsky Reconsidered,* by the editors of *Pensée* (New York: Warner Books, 1977). From 1972 to 1974 ten issues of *Pensée* were devoted to the ideas of Velikovsky. The selection included here was first published in May 1972 and was incorporated as a preface to *Velikovsky Reconsidered* with minor changes.

4. Critical reviews may be found in *Popular Astronomy,* June 1950, by Cecilia Payne-Gaposchkin, and in *Harper's Magazine,* June 1951, by John Q. Stewart. Generally favorable analyses of Velikovsky's work may be found in the issues of *Pensée* and in *Velikovsky Reconsidered.*

5. Newton, Dalton, Count Rumford, Einstein, Watson, for example.

6. Thomas Kuhn, *The Structure of Scientific Revolutions,* 2d ed. (Chicago: University of Chicago Press, 1970).

7. In *Against Method* (New York: Schocken, 1978), Paul Feyerabend has taken a position in opposition to this one, however.

8. For further examples, refer to *Velikovsky Reconsidered.*

/3/ CAUSALITY AND EXPLANATION

Proposal: The Most Adequate Explanation Is the One That Identifies Causes

Learning about science is quite different from doing science. As we learn about it, science often appears to be a neatly constructed building, brick laid upon brick, with the appearance of each new body of data leading to agreed-upon generalizations and explanations of the data. A study of the history of science, wherein we try to relive the thinking of the greatest scientific pioneers, however, reveals little of this kind of tidiness. Scientific contemporaries often accept radically different explanations for the same phenomena. Following Priestley's discovery of oxygen, a generation of chemists put the discovery to use within the framework of Lavoisier's explanation of combustion, but Priestley himself steadfastly adhered to the explanation that involved reference to phlogiston.[1] Ptolemaic astronomy actively competed with the new Copernican astronomy before it was abandoned. Examples could easily be multiplied, showing that within the scientific community there may be considerable dispute concerning the relative adequacy of two competing explanations.

In Chapter 2 we looked for criteria that one would use to distinguish scientific explanations from those that are pseudoscientific. Yet even if we had settled that issue, we would still need guidelines to help us find the best explanation among those that although all recognized as scientific are nonetheless not equally adequate. No one seems to have dismissed Priestley as a crank in the way that some scientists have reacted to Velikovsky (though scientists have been known to ascribe stubbornness to their colleagues). A study of the concept of explanation, within science, must take into account the variety of explanations that recognized scientists work with and accept. Such a study must show how a more adequate explanation is chosen from those that are equally legitimate or scientific.

Although the question of how one explanation can be more adequate or better than another has grown complex in modern times, it is interesting to note that it was a fairly simple question for many centuries. There is evidence even in some of the earliest recorded history, as well as ample evidence in the age of classical Greece, that there was one overriding understanding of what explanation was. To *explain* was to *identify the cause of a phenomenon or an event*. One modern commentator on the history of the notion of causality has stressed its importance in very early societies:

> According to a very popular belief, causality typifies modern science from its beginning until the birth of quantum mechanics, that is, roughly from the middle of the sixteenth century to our day. But most philosophers, and some scientists, know that the causal principle has survived the birth of the quantum theory, and that causal thought is much older than modern science. Explanation by causes is, indeed, as old as the phenomenological description of sheer time sequences. Moreover, the reduction of determination to causation is found in rather backward stages of knowledge. . . . It seems, in fact, characteristic of primitive mentality, at least at a certain stage of its evolution, to assign a cause to everything that is, begins to be, or passes away, and, particularly, to invent myths for explaining causally the origin of what we now regard as self-existent, unengendered, uncaused, namely, the universe as a whole; thus many cosmogonies, whether religious or not, besides fulfilling a social function, satisfy the urge for causal explanations. A second typical characteristic of primitive mentality is the ignorance of chance, the refusal to believe in mere conjunctions and fortuitous coincidences, and the complementary belief that all events are causally connected, whether in an overt or in a hidden (magical) way. This belief in the universal causal interconnection—a belief probably born in prehistoric times—was adopted in antiquity by Stoicism and is nowadays held by the continuers of prehistoric thought.[2]

The urge for causal explanations is still with us. We continue to use both the word *cause* and the mode of inquiry it suggests in our everyday existence. And many modern thinkers continue to stress the importance and legitimacy of the search for causes as a scientific enterprise. The great nineteenth-century philosopher of science John Stuart Mill would still find considerable support today for his view that the causal principle is "a main pillar of the inductive sciences."[3] The popularity of causal explanation is deceptive, however, because it masks some historic disagreements about the very meaning of causality.

What, precisely, is the "causal principle"? What does one mean, when one says that "to explain an event is to identify its cause"? We would no doubt have wide agreement that a cause is something that invariably precedes its effect, either immediately or through a chain of intermediaries. But there is something further in the usual meaning of *cause*. Notice that in the course of our experience with new phenomena, we may not immediately want to claim that a given event is always and invariably preceded by another particular event. The fact that you have repeatedly seen your instructor walk into the room immediately after the bell rings does not lead you to believe that she always will or that the bell *must* ring before she appears. You know enough to avoid attributing necessity to the sequence. Causes, as we normally appeal to them, are distinctive because they seem to possess this added feature, the *necessity* that a cause will be followed by an effect. This feature is reflected in our belief that a cause contains the *power* to produce an effect. Thus we not only want to say that a cause is something that always precedes its effect and is always followed by its effect; we also want to account for these relationships by attributing some productive quality—some power—to the cause.[4] Day follows night, and the movement of earth follows the impact of a bulldozer blade. Yet the blade causes the earth movement, while night doesn't cause day in any ordinary sense. Our confidence in the bulldozer blade as a cause is buttressed by our awareness of its momentum and acceleration—of the force it possesses, which qualifies it as a cause. The necessary connection between cause and effect seems accounted for by the qualities possessed by the cause.

The causal principle holds that events can be seen as the effects of causes that must be both necessary and sufficient, and a system of such interconnected causes and effects is called a causal nexus. If the universe as a whole is seen as one such causal nexus, the events in it are said to be causally determined. One may also hold a given event to be causally determined, or a given system to be a causal nexus, without holding that the universe as a whole is so determined. Whether the universe as a whole is determined or whether isolated systems may be considered determined is part of a problem we will return to in Chapter 10.

Why identify explanation with the presenting of causes? For many, the answer lies in the notions of power and productive force mentioned above. It would seem that to locate the power behind the occurrences of phenomena is to "unlock the secrets of nature," to "find out what makes things work," or to "be able to control nature by manipulating the causes of things." This enterprise is appealing to most of us.

Types of Causal Explanations

Let us then develop in this chapter the idea that to explain an event is to give its causes and that a good scientific explanation, whatever else it may be, is at least one that strives with rigor to locate, better than have previous explanations, the causes for the events in which we are interested. In order to do this we will have to look at the concept of cause, which is itself complex. A little reflection will convince us that very different sorts of explanations might be meant by an appeal to causes. Later in this chapter we will examine attacks on the causal principle as the basis for scientific explanation and responses to these attacks.

One way to become convinced about the variety of causal questions is to consider the questions that children ask. Suppose your little brother has seen you playing pool. You size up your shot, hit the cue ball, and it hits the five ball, which proceeds to roll slowly at a 45-degree angle into the corner pocket. Your brother asks, "Why did that happen?" Now what is he asking? Perhaps you think of your brother as an inquisitive young geometer, and you proceed to explain that the directions taken by the cue ball and the five ball after impact were instances of a complex pattern of geometrical relations, which explains why the five ball didn't go straight ahead.

But suppose you offer this perfectly reasonable answer, and your brother still looks puzzled. You will then probably explore other ways to make clear what happened, especially since you are feeling patient, having made the shot. You try again. "Do you mean, 'Why didn't the five ball break in pieces instead of rolling into the corner pocket?'" You suspect here that your brother is not ready to ask the geometrical question but is instead asking a "material" question. You are prepared to give an answer concerning the materials out of which pool balls are constructed, materials that give them a little elasticity but also considerable strength; but you notice that this variety of causal answer hasn't removed the puzzlement from your brother's face either.

At this point, a third kind of causal explanation might strike you. Perhaps your brother was not concentrating on the game and did not see you hit the cue ball or see the cue ball impact with the five ball. Suspecting this, you might say, "This white ball hit the other ball, and that's why it moved." In this instance, you have appealed not to the pattern of the event, or to the materials, but to that which, by its movement, caused movement somewhere else.

But suppose that you are having a very poor day for communication, and your brother still appears baffled over the course of events. Notice that there is still another "causal" sense of the question "Why did this event happen the way it did?" that we have not yet entertained. Perhaps your brother is inquiring about the intent or purpose behind the ball's falling into the corner pocket. Why did you want it to go there, if you did? So you answer, "It went into the pocket because I knocked it there, and I did that to win the game. And if you don't stop asking questions I'll knock you into the same pocket."

One of the earliest thinkers to deal systematically with the variety of causal questions was Aristotle. As part of his analysis, Aristotle recognized all four of the foregoing senses in which causal questions might be asked and insisted that before we can be confident that we have achieved scientific knowledge of an event, we should at least try to answer all four. He named and defined these senses as follows:[5]

1. The *formal cause* is the "essence," or structure, shape, or form of the event or the object. Certainly the resort to the angles of impact would be an appeal to the formal cause.

2. The *material cause* is the underlying matter that by virtue of its qualities (of hardness and elasticity in the above case) may allow an event to take place in a particular way.

3. The *efficient cause* is that which imparts the motion in question ("the primary source of the change or the coming to rest"). In this case there is a chain of efficient causes, beginning with the cue ball and going back to the cue, and your arm, etc.

4. The *final cause* (or *teleological* cause) is "that for the sake of which a thing is done," which in this case, assuming that you intended to make the shot you did, would be victory in the game.

Clearly, all four causal questions are not always appropriate. One doesn't ask for a material cause to account for the properties of a triangle. Nor does anyone but an exceedingly suspicious prosecutor persist forever in assuming an intent, or final cause, behind every auto accident. Furthermore, these categories of causal questions are not meant to be mutually exclusive. We could develop difficult examples in which one would be unclear about which kind of cause was involved. But the list was thought by Aristotle to be *exhaustive*. The value of having an exhaustive list lies in thereby knowing all the kinds of questions that you need to ask if you are seeking scientific knowledge. Answering fully the four causal questions should satisfy a scientist's (or a little brother's) quest for an understanding of an occurrence.

Aristotle's classification of causes has continued to be an important framework for thinking about explanation. In fact, we can use his classification to understand better some of the major arguments in which scientists and philosophers of science have engaged. For the most part, these arguments involve various attempts to emphasize one type of causal explanation as especially illuminating, or to eliminate one or more types as unscientific, or to reduce some types to others. Let us examine a few such attempts.

The Emphasis on Formal Cause. Mathematics seemed to René Descartes to provide the ideal method for the sciences. Convinced, by his work in analytic geometry, that physics was best approached mathematically, he argued that emphasis on the *structure* of events should replace concern with their *purposes*. In effect, this recommendation amounted to an endorsement of formal causes in preference to final causes. Descartes believed that speculations concerning the "end of nature," the "design of events," the "purposes of God" were not knowable for physical things anyway, and were therefore a waste of time.

> . . . the species of cause termed final, finds no useful employment in physical [or natural] things; for it does not appear to me that I can without temerity seek to investigate the [inscrutable] ends of God.[6]

Descartes' point of view constituted an important part of the modern scientific revolution. With the increasing importance of mathematics has come continuing attention to structure, or form, in scientific explanation.

The Continuing Defense of Final Cause. Aristotle's treatment of final causation provided medieval scientific thinkers with a handy tool for merging science and theology. If use of concepts such as the "design of Nature" and the purpose of an event or regularity were accepted as legitimate parts of a scientific explanation, they were also the key to knowledge of God's intentions and the reasonableness of his creations. Descartes' arguments that theology and physics should be kept distinct might therefore be thought to have dealt a serious blow to the use of final causes in natural science. Yet they continued to be defended not just for their utility in merging science and theology but for their power in guiding scientific discovery. Before Descartes, Kepler had seen his researches as leading to final cause explanations. Kepler believed that he had "reached a new conception of causality; that is, he thought of the underlying mathematical harmony discoverable in the observed facts as the cause of the latter, the reason, as he usually put it, why they are as they are."[7]

Following Descartes, even those thinkers who agreed with him about the importance of formal cause explanations—the mathematical structure of events as a key to understanding them—often continued to demand recognition for final causes. Leibniz, the great German philosopher of the seventeenth century, argued that we ought have no qualms about appealing to reasons in nature. He cited Snell, as one who productively used appeal to final cause explanations.

> It seems to me that Snell, who was the first discoverer of the laws of re-
> fraction, would have waited a long time before finding them if he had
> wished to seek out first how light was formed. But he apparently followed
> that method which the ancients employed for Catoptrics, that is, the
> method of final causes. Because, while seeking for the easiest way in which
> to conduct a ray of light from one given point to another given point by
> reflection from a given plane (supposing that that was the design of nature)
> they discovered the equality of angles of incidence and reflection as can
> be seen from a little treatise from Heliodorus of Larissa and also
> elsewhere. . . . That demonstration of this same theorem which Descartes
> has given, using efficient causes, is much less satisfactory. At least we have
> grounds to think that he [Descartes] would never have found the principle
> by that means if he had not learned in Holland of the discovery of Snell.[8]

In recent times the appeal to final causes, the ends or purposes of nature, has become less and less acceptable to most members of the scientific community. Why? The most persuasive reason is that scientists have come to see as useless those explanations for which no test at all is appropriate. This lack of testability seems to characterize at least those broad appeals to the design of nature that can yield no predictions whatever.

There are, however, two ways in which final cause explanations may still be useful. First, physics since the time of Newton has come to rely heavily on the notion of force. Yet Descartes recommended that the notion of force should not play a part in physics, because he saw it as a surreptitious appeal to final causation. To use forces, he thought, was to attribute to nature a design, whereby particles, for example, "naturally tend" to one pole of a magnet. Such terms as *attraction* and *repulsion* would have seemed to him giveaways that intentions were being appealed to where they had no place. Of course, most contemporary physicists would deny that they refer either to the purposes of nature or the intentions of the magnet when they use a word like *attraction*. Before the appeal to final causation is rejected as a legitimate mode of scientific explanation, one will at least need to consider carefully whether such concepts as "force" need to be or can be freed from finalistic overtones.

Final cause or teleological explanation may also be of relevance in the field of biology. Though scientists resist attributing intentions to inanimate nature or to individual atoms references are still sometimes made to the intentions of organisms and even species. Why does a spider spin its web? Surely the most direct answer, if your persistent little brother is asking the question, is "in order to catch insects." The language used in this answer seems clearly teleological, or finalistic. We refer, though perhaps metaphorically, to the spider's intent. Why did male peacocks develop elaborate plumage? To attract females? Notice that though we understand this purported explanation well enough, its simplicity is deceptive. An individual peacock doesn't in any sense choose its plumage. It is a genetically determined character of the species. But does a species choose, even metaphorically?

Among those biologists who agree to reject a literal attribution of real intentions to species and individual organisms, there has been dispute over whether even the metaphor is necessary or useful. After all, there does seem to be a design that fits the animal behavior to the best interests of the species. One respected contemporary biologist, Ernst Mayr, has attempted to show where purposiveness may be appropriately used in science.

Where, then, is it legitimate to speak of purposes and purposiveness in nature, and where is it not? To this question we can now give a firm and unambiguous answer. An individual who—to use the language of the computer—has been "programmed" can act purposefully. Historical processes, however, can *not* act purposefully. A bird that starts its migration, an insect that selects its host plant, an animal that avoids a predator, a male that displays to a female—they all act purposefully because they have been programmed to do so. When I speak of the programmed "individual," I do so in a broad sense. A programmed computer itself is an "individual" in this sense, but so is, during reproduction, a pair of birds whose instinctive and learned actions and interactions obey, so to speak, a single program.

The completely individualistic and yet also species-specific DNA code of every zygote (fertilized egg cell), which controls the development of the central and peripheral nervous systems, of the sense organs, of the hormones, of physiology and morphology, is the program for the behavior computer of this individual.

Natural selection does its best to favor the production of codes guaranteeing behavior that increases fitness. A behavior program that guarantees instantaneous correct reaction to a potential food source, to a potential enemy, or to a potential mate will certainly give greater fitness in the Darwinian sense than a program that lacks these properties. Again, a behavior program that allows for appropriate learning and the improvement of behavior reactions by various types of feedbacks gives greater likelihood of survival than a program that lacks these properties.

The purposive action of an individual, insofar as it is based on the properties of its genetic code, therefore is no more or less purposive than the actions of a computer that has been programmed to respond appropriately to various inputs.[9]

Mechanistic Science: The Appeal to Efficient Causes. In modern science there is a continuing emphasis on formal cause explanations, especially in the application of mathematics in science, and a continuing debate on the role, if any, of final cause. But the chief characteristic of modern science consists in its appeal to efficient cause explanations. The emphasis on mechanical push-pull models of causation began as early as Kepler, Galileo, and Hobbes. Newton expressed dissatisfaction with overreliance on Descartes' recommendation of formal systems as the key to scientific explanation. The triumph of classical mechanics in the eighteenth and nineteenth centuries seemed for a time to make mechanical explanations the ideal for all of science. Although contemporary physics also uses other models along side the mechanical one,[10] efficient cause explanations remain an important ideal in several branches of science. One commentator, Mario Bunge, sees the attempt to reduce the kinds of explanation to efficient, mechanical causal principles as a decisive characteristic of modern science.[11]

Attack on the Notion of Cause

We have seen that, for many people through many centuries, to explain a phenomenon scientifically consisted in identifying its causes. The great questions did, and to some extent still do, concern only the kind of cause that is the most appropriate object of the scientist's search. Yet one of the most startling developments in the philosophy of science over the past three centuries has been the attack on the notion of causality itself. More precisely, many thinkers have come to doubt our ability to know the real causes of phenomena and have therefore advocated that scientists either minimize or abandon the use of the notion of cause.

The most forceful critique of causal knowledge came in the work of the British empiricist David Hume. His philosophy represented a culmination of the development of empiricist thought in Great Britain in the seventeenth and eighteenth centuries. The basic claim of the empiricist school was outlined earlier—in 1690—by John Locke in his *Essay Concerning Human Understanding.* First, empiricists insisted that we restrict our claims about reality to those that we are competent to make. Locke called this a move toward critical philosophy, by which he meant that we ought to search out the limits to our possible understanding before we make farfetched claims that are impossible to verify.

What, then, for the empiricists, was the limit of knowledge? They answered that to find the limit, we ought to trace the sources of our knowledge, and that if we do so we will find that all of our knowledge comes from experience. It follows that if one cannot give a truth claim a "pedigree" by citing the evidence for it in experience, one ought not make such a claim at all.

By the time of Hume's writing, this empiricist doctrine had become fairly specific. Hume's general program was to seek out the sources of all of our ideas *in the impressions of which those ideas are simply copies* and then to allow no claims to be made about reality that could not be gotten from the original impressions. Hume wanted to locate in our experience the source of the idea of cause. The following selection is taken from Hume's *Inquiry Concerning Human Understanding*, published in 1748.[12]

OF THE IDEA OF NECESSARY CONNEXION

PART I.

There are no ideas which occur in metaphysics more obscure and uncertain than those of power, force, energy, or necessary connexion, of which it is every moment necessary for us to treat in all our disquisitions. We shall, therefore, endeavor, in this section, to fix, if possible, the precise meaning of these terms and thereby remove some part of that obscurity which is so much complained of in this species of philosophy.

It seems a proposition, which will not admit of much dispute, that all our ideas are nothing but copies of our impressions or, in other words, that it is impossible for us to think of anything which we have not antecedently felt, either by our external or internal senses. I have endeavoured to explain and prove this proposition and have expressed my hopes that, by a proper application of it, men may reach a greater clearness and precision in philosophical reasonings than what they have hitherto been able to attain. . . .

To be fully acquainted, therefore, with the idea of power or necessary connexion, let us examine its impression; and in order to find the impression with greater certainty, let us search for it in all the sources from which it may possibly be derived.

When we look about us towards external objects and consider the operation of causes, we are never able, in a single instance, to discover any power or necessary connexion, any quality which binds the effect to the cause and renders the one an infallible consequence of the other. We only find that the one does actually, in fact, follow the other. The impulse of one billiard ball is attended with motion in the second. This is the whole that appears to the outward senses. The mind feels no sentiment or inward impression from this succession of objects: Consequently, there is not, in any single, particular instance of cause and effect, anything which can suggest the idea of power or necessary connexion.

From the first appearance of an object, we never can conjecture what effect will result from it. But were the power or energy of any cause discoverable by the mind, we could foresee the effect, even without experience, and might, at first, pronounce with certainty concerning it, by the mere dint of thought and reasoning.

In reality, there is no part of matter that does ever by its sensible qualities discover any power or energy or give us ground to imagine that it could produce anything or be followed by any other object, which we could denominate its effect. Solidity, extension, motion; these qualities are all complete in themselves, and never point out any other event which may result from them. The scenes of the universe are continually shifting, and one object follows another in an uninterrupted succession; but the power or force which actuates the whole machine is entirely concealed from us and never discovers itself in any of the sensible qualities of body. We know that, in fact, heat is a constant *attendant of flame; but what is the connexion between them, we have no room so much as to conjecture or imagine. It is impossible, therefore, that the idea of power can be derived from the contemplation of bodies, in single instances of their operation because no bodies ever discover any power which can be the original of this idea.*

Since, therefore, external objects as they appear to the senses give us no idea of power or necessary connexion by their operation in particular instances, let us see whether this idea be derived from reflection on the operations of our own minds, and be copied from any internal impression. It may be said that we are every moment conscious of internal power, while we feel that, by the simple command of our will, we can move the organs of our body or direct the faculties of our mind. An act of volition produces motion in our limbs or raises a new idea in our imagination. This influence of the will we know by consciousness. Hence we acquire the idea of power or energy and are certain that we ourselves and all other intelligent beings are possessed of power. This idea, then, is an idea of reflection since it arises from reflecting on the operations of our own mind and on the command which is exercised by will, both over the organs of the body and faculties of the soul.

We shall proceed to examine this pretension, and first with regard to the influence of volition over the organs of the body. This influence, we may observe, is a fact which, like all other natural events, can be known only by experience and can never be foreseen from any apparent energy or power in the cause which connects it with the effect and renders the one an infallible consequence of the other. The motion of our body follows upon the command of our will. Of this we are every moment conscious. But the means by which this is effected, the energy by which the will performs so extraordinary an operation, of this we are so far from being immediately conscious that it must forever escape our most diligent inquiry.

For first, is there any principle in all nature more mysterious than the union of soul with body, by which a supposed spiritual substance acquires such an influence over a material one that the most refined thought is able to actuate the grossest matter? Were we empowered, by a secret wish, to remove mountains or control the planets in their orbit, this extensive authority would not be more extraordinary nor more beyond our comprehension. But if by consciousness we perceived any power or energy in the will, we must know this power; we must know its connexion with the effect; we must know the secret union of soul and body, and the nature of both these substances by which the one is able to operate, in so many instances, upon the other.

Secondly, we are not able to move all the organs of the body with a like authority, though we cannot assign any reason besides experience for so remarkable a difference between one and the other. Why has the will an influence over the tongue and fingers, not over the heart or liver? This question would never embarrass us were we conscious of a power in the former case, not in the latter. We should then perceive, independent of experience, why the authority of will over the organs of the body is circumscribed within such particular limits. Being in that case fully acquainted with the power or force by which it operates, we should also know, why its influence reaches precisely to such boundaries and no farther.

A man suddenly struck with a palsy in the leg or arm or who had newly lost those members frequently endeavors, at first, to move them and employ them in their usual offices. Here he is as much conscious of power to command such limbs as a man in perfect health is conscious of power to actuate any member which remains in its natural state and condition. But consciousness never deceives. Consequently, neither in the one case nor in the other are we ever conscious of any power. We learn the influence of our will from experience alone. And experience only teaches us how one event constantly follows another, without instructing us in the secret connexion which binds them together and renders them inseparable.

Thirdly, we learn from anatomy that the immediate object of power in voluntary motion is not the member itself which is moved but certain muscles, and nerves, and animal spirits, and, perhaps, something still more minute and more unknown through which the motion is successively propagated ere it reach the member itself whose motion is the immediate object of volition. Can there be a more certain proof that the power by which this whole operation is performed, so far from being directly and fully known by an inward sentiment or consciousness, is, to the last degree, mysterious and unintelligible? Here the mind wills a certain event. Immediately another event, unknown to ourselves, and totally different from the one intended, is produced. This event produces another, equally unknown. Till at last, through a long succession, the desired event is produced. But if the original power were felt, it must be known: Were it known, its effect must also be known since all power is relative to its effect. And vice versa, if the effect be not known, the power cannot be known or felt. How indeed can we be conscious of a power to move our limbs when we have no such power; but only that to move certain animal spirits, which, though they produce at last the motion of our limbs, yet operate in such a manner as is wholly beyond our comprehension?

We may, therefore, conclude from the whole, I hope, without any temerity though with assurance, that our idea of power is not copied from any sentiment or consciousness of power within ourselves when we give rise to animal motion or apply our limbs to their proper use and office. That their motion follows the command of the will is a matter of common experience, like other natural events. But the power or energy by which this is effected, like that in other natural events, is unknown and inconceivable. . . .

The generality of mankind never find any difficulty in accounting for the more common and familiar operations of nature, such as the descent of heavy bodies, the growth of plants, the generation of animals, or the nourishment of bodies by food. But suppose, that, in all these cases, they perceive the very force or energy of the cause by which it is connected with its effect and is forever infallible in its operation. They acquire, by long habit, such a turn of mind, that, upon the appearance of the cause, they immediately expect with assurance its usual attendant, and hardly conceive it possible that any other event could result from it. It is only on the discovery of extraordinary phenomena, such as earthquakes, pestilence, and prodigies of any kind, that they find themselves at a loss to assign a proper cause and to explain the manner in which the effect is produced by it. It is usual for men in such difficulties to have resource to some invisible intelligent principle as the immediate cause of that event which surprises them and which they think cannot be accounted for from the common powers of nature. But philosophers, who carry their scrutiny a little farther, immediately perceive, that even in the most familiar events the energy of the cause is as unintelligible as in the most unusual, and that we only learn by experience the frequent CONJUNCTION of objects, without being ever able to comprehend anything like CONNEXION between them. . . .

PART II.

But to hasten to a conclusion of this argument, which is already drawn out to too great a length: We have sought in vain for an idea of power or necessary connexion in all the sources from which we could suppose it to be derived. It appears, that in single instances of the operation of bodies we never can, by our utmost scrutiny, discover anything but one event following another, without being able to comprehend any force or power by which the cause operates or any connexion between it and its supposed effect. The same difficulty occurs in contemplating the operations of mind on body, where we observe the motion of the latter to follow upon the volition of the former but are not able to observe or conceive the tie which binds together the motion and volition, or the energy by which the mind produces this effect. The authority of the will over its own faculties and ideas is not a whit more comprehensible: So that, upon the whole, there appears not throughout all nature any one instance of connexion which is conceivable by us. All events seem entirely loose and separate. One event follows another; but we never can observe any tie between them. They seem conjoined but never connected. And as we can have no idea of anything which never appeared to our outward sense or inward sentiment, the necessary conclusion seems to be that we have no idea of connexion or power at all and that these words are absolutely without any meaning when employed either in philosophical reasonings or common life.

But there still remains one method of avoiding this conclusion and one source which we have not yet examined. When any natural object or event is presented, it is impossible for us, by any sagacity or penetration, to discover, or even conjecture, without experience, what event will result from it or to carry our foresight beyond that object which is immediately present to the memory and senses. Even after one instance or experiment where we have observed a particular event to follow upon another, we are not entitled to form a general rule or foretell what will happen in like cases, it being justly esteemed an unpardonable temerity to judge of the whole course of nature from one single experiment, however accurate or certain. But when one particular species of event has always in all instances been conjoined with another, we make no longer any scruple of foretelling one upon the appearance of the other and of employing that reasoning which can alone assure us of any matter of fact or existence. We then call the one object Cause; the other, Effect. We suppose, that there is some connexion between them; some power in the one, by which it infallibly produces the other, and operates with the greatest certainty and strongest necessity.

It appears, then, that this idea of a necessary connexion among events arises from a number of similar instances which occur of the constant conjunction of these events; nor can that idea ever be suggested by any one of these instances surveyed in all possible lights and positions. But there is nothing in a number of instances different from every single instance which is supposed to be exactly similar except only that after a repetition of similar instances, the mind is carried by habit, upon the appearance of one event, to expect its usual attendant and to believe that it will exist. This connexion, therefore, which we feel in the mind, this customary transition of the imagination from one object to its usual attendant, is the sentiment or impression from which we form the idea of power or necessary connexion. Nothing farther is in the case. Contemplate the subject on all sides; you will never find any other origin of that idea. This is the sole difference between one instance, from which we can never receive the idea of connexion, and a number of similar instances by which it is suggested. The first time a man saw the communication of motion by impulse, as by the shock of two billiard balls, he could not pronounce that the one event was connected but only that it was conjoined with the other. After he had observed several instances of this nature, he then pronounces them to be connected. What alteration has happened to give rise to this new idea of connexion? Nothing but that he now feels these events to be connected in his imagination and can readily foretell the existence of one from the appearance of the other. When we say, therefore, that one object is connected with another, we mean only that they have acquired a connexion in our thought and give rise to this inference by which they become proofs of each other's existence: A conclusion which is somewhat extraordinary but which seems founded on sufficient evidence. Nor will its evidence be weakened by any general diffidence of the understanding or sceptical suspicion concerning every conclusion which is new and extraordinary. No conclusions can be more agreeable to scepticism than such as make discoveries concerning the weakness and narrow limits of human reason and capacity.

And what stronger instance can be produced of the surprising ignorance and weakness of the understanding than the present? For surely, if there be any relation among objects which it imports to us to know perfectly, it is that of cause and effect. On this are founded all our reasonings concerning matter of fact or existence. By means of it alone we attain any assurance concerning objects which are removed from the present testimony of our memory and senses. The only immediate utility of all sciences is to teach us how to control and regulate future events by their causes. Our thoughts and enquiries are, therefore, every moment employed about this relation. Yet so imperfect are the ideas which we form concerning it that it is impossible to give any just definition of cause except what is drawn from something extraneous and foreign to it. Similar objects are always conjoined with similar. Of this we have experience. Suitably to this experience, therefore, we may define a cause to be an object, followed by another, and where all the objects similar to the first are followed by objects similar to the second. Or, in other words, where, if the first object had not been, the second never had existed. The appearance of a cause always conveys the mind, by a customary transition, to the idea of the effect. Of this also we have experience. We may, therefore, suitably to this experience, form another definition of cause and call it an object followed by another, and whose appearance always conveys the thought to that other. But though both these definitions be drawn from circumstances foreign to the cause, we cannot remedy this inconvenience or attain any more perfect definition which may point out that circumstance in the cause which gives it a connexion with its effect. We have no idea of this connexion or even any distinct notion what it is we desire to know when we endeavor at a conception of it. We say, for instance, that the vibration of this string is the cause of this particular sound. But what do we mean by that affirmation? We either mean that this vibration is followed by this sound, and that all similar vibrations have been followed by similar sounds; or that this vibration is followed by this sound, and that upon the appearance of one, the mind anticipates the senses and forms immediately an idea of the other. *We may consider the relation of cause and effect in either of these two lights; but beyond these, we have no idea of it.*

To recapitulate, therefore, the reasonings of this section: Every idea is copied from some preceding impression or sentiment; and where we cannot find any impression, we may be certain that there is no idea. In all single instances of the operation of bodies or minds, there is nothing that produces any impression, nor consequently can suggest any idea, of power or necessary connexion. But when many uniform instances appear, and the same object is always followed by the same event, we then begin to entertain the notion of cause and connexion. We then feel a new sentiment or impression, to wit, a customary connexion in the thought or imagination between one object and its usual attendant; and this sentiment is the original of that idea which we seek for. For as this idea arises from a number of similar instances and not from any single instance, it must arise from that circumstance in which the number of instances differ from every individual instance. But this customary connexion or transition of the imagination is the only circumstance in which they differ. In every other particular, they are alike. The first instance which we saw of motion communicated by the shock of two billiard balls (to return to this obvious illustration) is exactly similar to any instance that may, at present, occur to us, except only that we could not, at first, infer one event from the other, which we are enabled to do at present, after so long a course of uniform experience. . . .

Analysis of Hume's Argument

The Force of the Argument. Hume had reason to complain that his contemporaries paid little attention to his analysis of causality and to the skepticism, concerning the possibility of knowledge, that was implied by it. Yet, the passage of time has witnessed increasing attention to the force of Hume's analysis. His argument is all the more startling because he is not merely claiming that our knowledge of cause and effect is *limited*; rather, he is claiming that we have no knowledge of causes in the world, whatsoever. The serious nature of Hume's argument will appear more clearly if we can draw out some of its implications.[13] Hume believed that all of our knowledge comes from impressions made upon our senses or arises internally in us as feelings. What causes these impressions, from where they come, it is impossible for us to ever know because we cannot get "behind" them. As to the connections between impressions, all that we are aware of in experience is the past sequence in which some object *a* has been followed by an object *b*. No matter how many times the sequence occurs in the same way, however, we can never find in our experience of the sequence itself any "agency" or ongoing "power" in the first object that would assure us of its operating the same way in the future. Therefore, we have no knowledge of causality, and our belief that the future will be like the past is the sheerest presumption.

It would be a great mistake to conclude from Hume's argument that though we can never be *certain* that the future will be like the past or that a given object is indeed the cause of another object, we can still, given the weight of our past experience, anticipate the effect with some degree of reasonable probability. Hume has sometimes been interpreted this way, but if he is so interpreted, the full force of his argument is not appreciated. The reasonableness of probability is precisely what Hume was attacking. We have *no reason*, that is, no rational justification, for believing either that the future will be like the past in general or that this object that we have identified as a cause will continue to produce its particular effect.

Coherence and Correspondence. Hume declared the impossibility of our knowing the causes of things. Why, then, did he not recommend giving up the concept of causality? Why did he assure us that he himself would continue to make causal judgments?

To understand Hume's position we will need to recognize two competing theories of truth, one of which Hume rejected but the other of which he accepted.

Consider the two following propositions:

1. It sometimes snows in Canada.
2. In chess, the pawn can only attack diagonally.

Now, though we would hold both these propositions to be true, notice that we would normally justify them by appeal to different criteria. Isn't the first proposition true simply because it corresponds to a real state of affairs in the world? Presumably, it would sometimes snow in Canada even if no human had ever been there, or even if no human had ever existed. The truth of the proposition consists simply in its correspondence with the facts. The truth of proposition 2, however, seems to depend only on human thought. This is because it is consistent with, or "coheres" in, the system called "the rules of chess." It need not correspond with anything in the world, since even if all chessboards and chess pieces were destroyed, the rules of the game would not be changed.

Philosophers point out that for some propositions, such as number 2, coherence with other propositions is the criterion of truth or falsity. This is the criterion to which we appeal when we say that a poorly written fairy tale doesn't "make sense." We don't mean that it doesn't correspond to reality—fairy tales aren't supposed to! Rather, we mean that parts of the story don't "cohere" with the rest of the story.

There has, however, been considerable disagreement about whether any propositions are true because they correspond to a reality independent of human perception. The theory that there are propositions such as this is called the *correspondence* theory of truth. Proponents of this view insist that if our claims were not capable of reflecting a reality beyond human thought and experience, we could not claim to have any real knowledge at all. Opponents of the correspondence view of truth have usually subscribed to the *coherence* theory of truth. This theory denies that propositions are true because they correspond to a real state of affairs beyond human perceptions, and urges that propositions are true only in that they cohere with other propositions about our experience.

It is quite clear that those pioneers of modern science such as Descartes and Kepler believed that their resort to causal explanations was scientifically worthwhile because real causes exist in the world and these are knowable by the human mind. They believed, each with his own emphases and modifications, in the correspondence theory of truth. The ideal is for the relationship of ideas in our minds to reflect or mirror the real objects and events that exist independently of our knowing about them.

Hume asked us to reflect carefully on the justification for holding that our ideas are mirrors of a reality independent of human perception. He believed that we can trace ideas only as far as those sense impressions of which the ideas are themselves copies. But it then follows that the impressions themselves can be traced no further. We can know the relationship of those impressions in our minds, and we can organize coherent systems by which we can communicate about the relationship of ideas, but there is no way in the world that we can prove, or even adduce evidence, that there is a particular kind of reality "out there somewhere" of which our impressions are a reflection. Our attempt to acquire knowledge must then be an enterprise of relating ideas to one another rather than an attempt to hook up our ideas with real things. Causality thus became for Hume simply a relation of ideas having no knowable correspondence with a reality behind our impressions.

We can now understand why Hume was so anxious to evaluate the sources of our belief in causes. If you ask how someone knows that a given event or object is the cause of another event or object, and he or she can only reply that the two have gone together in our previous experience, you may still be willing to label the first of the two associated events as a cause. But Hume urged us not to convert that labeling decision, or convention, into an unsupportable claim about what's really going on in nature. Preceding Hume, philosophers had struggled with this difficulty and had sometimes argued that we are directly aware of the workings of nature in our consciousness of the reality of power. If we cannot see the causal power in external objects, at least we are aware of such power in our own will when we decide to raise our arm and then raise it. Notice that Hume took great pains with this argument because it concerns the core of his thesis. Taking each of the alleged accesses to reality that have been offered, he argued that we don't really have any access at all.[14]

Hume, then, considered the human employment of the notion of cause as perfectly natural, indeed as a part of the nature of human thinking. He did not recommend abandoning the concept but came to the skeptical conclusion that we have no evidence and could not in principle have any evidence that our presumptions concerning causality are rooted in a reality behind our impressions.

Earlier in this chapter we argued that much of the history of Western science, including the opinions of some of the great founders of modern science about what they were doing, can be seen as an attempt to explain by locating causes or by redefining what sort of causes science ought to be seeking as explanations. Now, however, we are faced with a critic who claims that scientists may make causal statements but must not delude themselves into believing that they have thereby gotten access to the world beyond human perceptions. Science may continue to systematize statements about the way humans perceive and reflect. However, that is all it can do.[15]

Responses to Hume's Argument

How ought we to respond to Hume's position? Should we give up the ideal of explanation as the location of real causes in nature? Or could it be that he made a mistake, either in his assumptions or in his reasoning, so that we may refute his position and maintain our confidence in explanation as cause identifying?

Since Hume's time, most philosophers of science have seen his position as important enough to demand a response. No single, agreed-upon position, either in defense of or rebuttal to Hume has emerged, however. For the sake of simplicity we may cluster the modern responses around four basic positions.

Rejection of Hume's Position. For over a century after Hume's work, the most popular attitude of professional philosophers toward it was to argue that the skepticism toward which it took us was evidence that the whole empiricist tradition was headed in the wrong direction. Those opposed to the empiricists, usually called *rationalists* in the history of philosophy, pointed to the rapid successes of the new science. They pointed out that we have, through the use of our reason, unlocked many of the secrets of the structure of nature. Therefore, skepticism concerning our ability to know the world as it is in itself hardly seems justified.

Although a distaste for skepticism provides some understanding of the opposition to Hume, it hardly by itself constitutes a philosophical argument. The points at which Hume's opponents have attacked his analysis are centered around two main issues. They are, first, the disagreement over whether or not we have ideas that could not possibly have come from experience (and thus could not be explained by Hume's theory of knowledge) and, second, the argument concerning whether we do receive our sense impressions in bits and pieces, as Hume seemed to believe. In pursuit of the first issue, the rationalists (from Professor Thomas Reid, Hume's contemporary at Edinburgh, to such thinkers as Brand Blanshard in this century) have argued that many of our ideas cannot be traced to sense experience, as Hume claimed. Rationalists often cite such examples as the concepts of infinity, matter, number, or equality. They argue that these ideas, wherever they come from, cannot be gotten even by abstraction from our experience; therefore, to try to disqualify any kind of knowledge simply by asserting that we are not justified in holding it on the basis of sense experience is unfair. The dispute over this issue goes back as far as Plato and Aristotle in ancient Greece and, with refinements, is still with us.

As nice as it would be to be able to conclude one way or another at this point in our discussion concerning the validity of either Hume's position or the antiempiricist position, to do so we would have to ignore the continuing divisions of opinion on the matter. It should be clear, however, that a great deal rests on the position one takes in this debate. Hume's analysis certainly does seem to depend upon the fundamental premise that all knowledge must come from experience. If that premise is false, then we need not worry about Hume's criticism of causal claims about the real world (though there might be other reasons for worrying about such claims).

A second traditional criticism of Hume concerns his theory of perception. His argument concludes that the necessary connection between two events (one alleged to be the cause of the other) must be only a presumption because, in our experience, these events or objects or fleeting phenomena are separate and distinct. In other words, when we open our eyes we see an array of colors and shapes, none of which are necessarily (logically) connected with any others. When we see the different parts of a chair, we see different shades of color as the sun plays upon the back and the seat and the arms. Hume believed we are actually seeing an array of separate impressions, which through their habitual connection in our experience we, from habit, clump together. We then pretend that they are merely different parts of one object, the chair. Hume believed that the object results from an act of our imagination rather than being something that we actually see. But what if it is the case that we do not in any meaningful sense "see" the separate sense impressions but rather learn, when we are children, what to look for by the names of objects taught to us by our parents? Do we really *see* things that we do not pay attention to, that we have no words for, and that we do not understand? Hume's opponents sometimes argue that since we "see" a connected field rather than individual and discreet bits and pieces Hume was quite wrong in insisting that we cannot claim to perceive connections. If Hume was wrong on this point, it becomes possible to argue that we directly sense causal connections. At least it is no longer necessary to deny that we have any evidence for them because of the separability and discreteness of our individual sense perceptions.[16]

The Kantian Response to Hume. While some might argue that Hume need not be a source of concern to philosophers of science because his doctrine is fundamentally wrong, other thinkers have believed that the justice of at least some of Hume's points calls for major modifications in our view of scientific explanation. Some of them, following Immanuel Kant, have accepted Hume's position that causality is not an appropriate area for *ontological* claims (that is, claims about what the world is like in itself and independent of human perceptions of it), but then go on and argue that this does not lead to the skeptical conclusions that Hume drew. Kant, in the *Critique of Pure Reason*,[17] argued that in order to accommodate the truth involved in Hume's analysis, we would have to redefine what we mean by experience and "the world." While freely admitting that claims we make about the world must, if they are to be reasonable, restrict themselves to our experience, Kant pointed out that the human mind does not experience discrete and separate units of perception and *then* manipulate the experience to form relations among the perceptions. Rather, he argued, by the time we have experience, the mind has already influenced the manifold of sensory impressions, such that our "world" is a world already inherently affected by human categories of organization. It therefore makes sense to say that "the world is causally connected," though admitting that the world of which we speak is *the world of human experience*, not some world as it exists in itself and independent of human experience. Kant then went on to argue that there are identifiable and absolutely universal ways in which human beings experience that world and therefore that we can retain the most important feature of scientific knowledge, namely, its intersubjectivity and lack of dependence upon opinion or "feelings" of any individual group. In other words, Kant willingly gave up the extensiveness of the ontological claims of causality but concluded that we really do not need to make that kind of claim anyway; knowledge of a world independent of human perception is not necessary in order to save us from an utter relativity and subjectivity that would make science as we know it impossible.

Deemphasis of Causality. Both Kant and the rationalists were concerned to preserve the integrity of appeals to causes. Though rationalism encompasses a varied group of thinkers, most of them would agree to accept causal explanation as an important, if not the only, kind of scientific explanation. However, many philosophers of science disagree. Generally speaking, those who have accepted the empiricists' position that all of our knowledge comes from experience tend either to deemphasize or reject entirely the search for causes as the appropriate model for explanation. Some philosophers of science, typified by Stephen Toulmin,[18] see the belief in a system of causes as reliance on a myth. Toulmin shares Hume's dissatisfaction with the appeal to causes as an appropriate ontological move. However, Toulmin respects the use of the concept of cause in science, particularly the applied sciences, as a diagnostic tool instead of a factual claim about the structure of the world. He believes this tool is most effective when we want to manipulate the outcome of events, and he therefore sees causality primarily as a directional guide for our research, or a prescription (as opposed to a description), or as an overarching methodological assumption that we do not pretend either to have evidence for or to be able to falsify. Toulmin's position is a moderate one. Like Hume, he criticizes the ontological pretentions of those who would *explain* by the identification of causes but is sympathetic to the continued use of the concept anyway.

Some others suggest that we simply abandon the concept of causality in science altogether. Recall that Ernst Mayr attempted to throw out the undesirable uses of *teleology* but to retain a remnant of its meaning worthwhile for science. In a similar but more sweeping move, these philosophers of science attempt to excise *cause* from the scientific vocabulary but to retain whatever was useful in the meaning of that term. Typical of such thinkers was Bertrand Russell:

> The law of causality, I believe, like much that passes muster among philosophers, is a relic of a bygone age, surviving, like the monarchy, only because it is erroneously supposed to do no harm.[19]

Russell wanted to acknowledge that something called the "uniformity of nature" is accepted by science, "on inductive grounds." He believed this acceptance of a uniformity of nature saves what was needed in the old "law of causality." It must be stressed that Russell was assuming the legitimacy of the principle of induction, that is, the principle that we can assign a probability to a future event on the basis of experience in the past. Hume would accept this principle only as a presumption and Russell did not undertake to define it as something more than a presumption in his work. It would of course be expected that those rationalists who rejected Hume's skepticism would quickly argue with Russell's position. They would insist that the only reason that one can be justified in a faith in the uniformity of nature is if one has some prior assurance about reliability of causes and the powers of their operation.[20]

Positivism and the Rejection of Explanation. One element is common to all of the foregoing responses to Hume's argument. Each point of view cited respects the notion that science is supposed to explain our experience. The rationalists generally continue to identify such explaining with the location of causes. Kantians will also continue to respect the power of causal judgments as components of explanation. Toulmin, Russell, and others who seek to deemphasize the traditional notion of causality would divorce it from explanation without abandoning the concept of explanation itself. But if causal judgments about the world can in principle have no evidence for them, as Hume suggested, is it not possible that this weakness is shared by any other kinds of purported explanation, causal or not? Perhaps we ought to take most seriously the general proposition that all we know we learn from experience, and couple it with the judgment that experience always tells us *that* something is the case rather than how or why it is the case. This beginning leads to the intriguing possibility that perhaps explanation ought to mean nothing more than description. Such a proposal has the merit of promising to avoid the tangles into which the consideration of causality has led us. In Chapter 4 we will consider the proposal of the positivist school in the philosophy of science, which asks us to abandon the attempt to define explanation either as causal or in any other way that takes science beyond the enterprise of merely describing what occurs in our experience.

Supplementary Readings

Pirsig, Robert. *Zen and the Art of Motorcycle Maintenance.* New York: Morrow, 1974, pp. 123–130.

A very readable version of the Humean argument and Kant's response to it, in the context of a novel.

Losee, John. *A Historical Introduction to the Philosophy of Science.* London: Oxford University Press, 1972.

A good source for discussion of work done by the ancient Greeks, Bacon, and some nineteenth-century philosophers of science to whom we do not give much space. Reading Losee will provide some perspective on the choices made in this book.

Oldroyd, David. *The Arch of Knowledge*. New York: Methuen, 1986.

Although nearly half of this work is devoted to twentieth-century philosophy of science, which is the subject of our later chapters, we suggest your trying the Oldroyd book as an alternative account as we move along. A good bibliography appears on pp. 373–383.

Copleston, Frederick. *A History of Philosophy*. Vol. 6. London: Burns, Oates and Washbourne Limited, 1964, pp. 235–276.

This is a standard work of great value to those who want to get a good view of a philosopher's ideas and place in history. Kant's importance is underscored by the fact that he is granted half of volume 6 in Copleston's eight-volume work.

Notes

1. For accounts of this controversy, see J. B. Conant, *Harvard Case Histories in Experimental Science* (Cambridge, Mass: Harvard University Press, 1950); and I. Freund, *The Study of Chemical Composition* (New York: Dover, 1968). The oversimplified version of this controversy, that Priestley was wrong and Lavoisier was right, has been rejected by Thomas Kuhn.

2. Mario Bunge, *Causality* (Cambridge, Mass: Harvard University Press, 1959), p. 224.

3. John Stuart Mill, *A System of Logic*, in *Selected Works of Mill*, vol. 7 (Toronto: University of Toronto Press, 1973), p. 327.

4. Notice that these are two different matters. To say that a cause always precedes its effect means that it is *necessary*. To say that it is always followed by its effect means that it is *sufficient*.

5. Aristotle, *Physics*, in *The Basic Works of Aristotle*, trans. R. McKeon (New York: Random House, 1941), pp. 240–241. Aristotle's words, as translated by McKeon, are in quotation marks.

6. René Descartes, *Meditations on First Philosophy*, vol. 4, *The Philosophical Works of Descartes*, trans. E. S. Haldane and G. R. T. Ross (London: Cambridge University Press, 1911).

7. Edwin A. Burtt, *The Metaphysical Foundations of Modern Physical Science* (New York: Doubleday, 1932), p. 53, quoted in Bunge, *Causality*, p. 228.

8. Leibniz, *Discourse on Metaphysics*, in *Leibniz, Selections* (New York: Scribner's, 1951), p. 323.

9. Ernst Mayr, "Cause and Effect in Biology," *Science* 134 (1961): 1503–1504. For a helpful discussion of teleology and mechanism, see Ian Barbour, *Issues in Science and Religion* (New York, Harper & Row, 1971), p. 337.

10. Quantum theory, for instance, is not currently confined by any insis-

tence that its ultimate units be particles of finite size and discrete spatial location.

11. Bunge, *Causality*, p. 226.

12. David Hume, "An Inquiry Concerning Human Understanding," in David Hume, *Philosophical Works*, vol. 3, ed. T. H. Green and T. H. Gross, London, 1882. Footnotes have been omitted.

13. Hume himself spelled out the implications of his analysis more fully in an earlier work, *A Treatise of Human Nature* (London, 1739).

14. It is ironic that the empiricists in their emphasis on empirical observations find themselves in the end cut off from an external reality, according to Hume's argument.

15. Some commentators on Hume believe that he did not hold to this skepticism consistently. For example, see R. F. Anderson, *Hume's First Principles* (Lincoln: University of Nebraska Press, 1966).

16. Whether Hume actually believed that we perceive perceptual "atoms" has been disputed by some Hume scholars. For a treatment of Hume's theory of perception that stresses its atomism, see H. H. Price, *Hume's Theory of the External World* (Oxford: Clarendon Press, 1963); for an argument opposing the view that Hume was a "perceptual atomist," see C. W. Hendel, *Studies in the Philosophy of David Hume* (Indianapolis: Bobbs-Merrill, 1963).

17. Immanuel Kant, *Critique of Pure Reason*, 1789, trans. Norman Kemp Smith (New York: St. Martin's Press, 1961).

18. Stephen Toulmin, *The Philosophy of Science* (London: Hutchinson University Library, 1953), p. 161.

19. Bertrand Russell, *Mysticism and Logic* (New York, W. W. Norton, 1929), p. 180.

20. We ought not leave this brief description of the noncausalist position without noting that many examples of explanations seem to be noncausal. An important further argument of this group of people is that they can develop translation procedures by which they can restate anything that someone thinks must be said causally into noncausal terms. For further discussion of this issue, see Mario Bunge, *Causality and Modern Science*, (New York: Dover, 1969), pp. 298ff.

/4/ EXPLANATION AND DESCRIPTION

Proposal: The Most Adequate Scientific Explanation Is Merely the Best Description

When we explain a phenomenon, are we merely describing it in a particular way, or are we doing something more than that? In Chapter 3 we assumed that explanation was something more than description. To do this we relied on the common distinction between saying that something is the case and saying why it is the case. It seemed natural to see explanation as the answer to the "why" questions.

Yet we have seen that attempts to characterize the "why" answers by referring to causality are burdened with difficulties. Empiricists caution us to make claims only about what we are acquainted with—our own experience. Hume argued that nowhere do we experience causality as anything more than a feeling. Responses to his arguments, at the very least, either involve us in complex philosophical disputes or force us to adopt the disagreeable consequences of Hume's position.

These difficulties alone might persuade us to seek another way to characterize scientific explanation. Would identifying explanation with description be so very far from the way we ordinarily use these terms? We would no doubt like to say that science "tells us what the world is like." That sounds like a descriptive enterprise. Furthermore, scientists appear to agree with one another on a great many things. Perhaps this is because they are simply engaged in reporting and organizing the facts. If we insist that explanation is a kind of description, we may do a better job of sticking to the main scientific task: revealing our world to us. We might better avoid the unanswerable speculative tangles that we get into by saying that explanation is something beyond description.

The above advantages suggest the descriptivist approach to explanation, which has been the central theme of a group of philosophers called the positivists. Positivism takes its name from, and finds its roots in, the publication of *Positive Philosophy*, by Auguste Comte, in 1830.[1] However, Comte's influence soon spread to many other thinkers in diverse fields. One of the most important later positivists was the physicist and philosopher of science Ernst Mach. Mach in turn became the guiding spirit behind the group of scientists and philosophers who formed the famous Vienna Circle in the early twentieth century. It is the work of this group of thinkers that, following Mach, gave positivism its classical formulation. One may find summaries of the positivist point of view in the works of many contemporary writers. Yet for clarity and pointedness, the words of Mach himself have a lasting quality that justifies their inclusion here. The following essay, edited for our purposes, is from the chapter on formal principles in Mach's *The Science of Mechanics* (1883). In reading Mach's analysis, keep in mind the evils that he was trying to avoid. How can we avoid mystical speculations in scientific work? To what clear basis in observation can we refer disputes?[2]

THE ECONOMY OF SCIENCE

1. *It is the object of science to replace, or save, experiences, by the reproduction and anticipation of facts in thought. Memory is handier than experience, and often answers the same purpose. This economical office of science, which fills its whole life, is apparent at first glance; and with its full recognition all mysticism in science disappears. Science is communicated by instruction, in order that one man may profit by the experience of another and be spared the trouble of accumulating it for himself; and thus, to spare posterity, the experiences of whole generations are stored up in libraries. . . .*

2. *In the reproduction of facts in thought, we never reproduce the facts in full, but only that side of them which is important to us, moved to this directly or indirectly by a practical interest. Our reproductions are invariably abstractions. Here again is an economical tendency.*

Nature is composed of sensations as its elements. Primitive man, however, first picks out certain compounds of these elements—those namely that are relatively permanent and of greater importance to him. The first and oldest words are names of "things." Even here, there is an abstractive process, an abstraction from the surroundings of the things, and from the continual small changes which these compound sensations undergo, which being practically unimportant are not noticed. No inalterable thing exists. The thing is an abstraction, the name a symbol, for a compound of elements from whose changes we abstract. The reason we assign a single word to a whole compound is that we need to suggest all the constituent sensations at once. When, later, we come to remark the changeableness, we cannot at the same time hold fast to the idea of the thing's permanence, unless we have recourse to the conception of a thing-in-itself, or other such like absurdity. Sensations are not signs of things; but, on the contrary, a thing is a thought-symbol for a compound sensation of relative fixedness. Properly speaking the world is not composed of "things" as its elements, but of colors, tones, pressures, spaces, times, in short what we ordinarily call individual sensations.

The whole operation is a mere affair of economy. In the reproduction of facts, we begin with the more durable and familiar compounds, and supplement these later with the unusual by way of corrections. Thus, we speak of a perforated cylinder, of a cube with beveled edges, expressions involving contradictions, unless we accept the view here taken. All judgments are such amplifications and corrections of ideas already admitted.

3. In speaking of cause and effect we arbitrarily give relief to those elements to whose connection we have to attend in the reproduction of a fact in the respect in which it is important to us. There is no cause nor effect in nature; nature has but an individual existence; nature simply is. Recurrences of like cases in which A is always connected with B, that is, like results under like circumstances, that is again, the essence of the connection of cause and effect, exist but in the abstraction which we perform for the purpose of mentally reproducing the facts. Let a fact become familiar, and we no longer require this putting into relief of its connecting marks, our attention is no longer attracted to the new and surprising, and we cease to speak of cause and effect. Heat is said to be the cause of the tension of steam; but when the phenomenon becomes familiar we think of the steam at once with the tension proper to its temperature. Acid is said to be the cause of the reddening of tincture of litmus; but later we think of the reddening as a property of the acid.

Hume first propounded the question: How can a thing A act on another thing B? Hume, in fact, rejects causality and recognizes only a wonted succession in time. Kant correctly remarked that a necessary connection between A and B could not be disclosed by simple observation. He assumes an innate idea or category of the mind, a Verstandesbegriff, under which the cases of experience are subsumed. Schopenhauer, who adopts substantially the same position, distinguishes four forms of the "principle of sufficient reason"—the logical, physical, and mathematical form, and the law of motivation. But these forms differ only as regards the matter to which they are applied, which may belong either to outward or inward experience.

The natural and commonsense explanation is apparently this. The ideas of cause and effect originally sprang from an endeavor to reproduce facts in thought. At first, the connection of A and B, of C and D, of E and F, and so forth, is regarded as familiar. But after a greater range of experience is acquired and a connection between M and N is observed, it often turns out that we recognize M as made up of A, C, E, and N of B, D, F, the connection of which was before a familiar fact and accordingly possesses with us a higher authority. This explains why a person of experience regards a new event with different eyes than the novice. The new experience is illuminated by the mass of old experience. As a fact, then, there really does exist in the mind an "idea" under which fresh experiences are subsumed; but that idea has itself been developed from experience. The notion of the necessity of the causal connection is probably created by our voluntary movements in the world and by the changes which these indirectly produce, as Hume supposed but Schopenhauer contested. Much of the authority of the ideas of cause and effect is due to the fact that they are developed instinctively and involuntarily, and that we are distinctly sensible of having personally contributed nothing to their formation. We may, indeed, say that our sense of causality is not acquired by the individual, but has been perfected in the development of the race. Cause and effect, therefore, are things of thought, having an economical office. It cannot be said why they arise. For it is precisely by the abstraction of uniformities that we know the question "why."

4. In the details of science, its economical character is still more apparent. The so-called descriptive sciences must chiefly remain content with reconstructing individual facts. Where it is possible, the common features of many facts are once for all placed in relief. But in sciences that are more highly developed, rules for the reconstruction of great numbers of facts may be embodied in a single expression. Thus, instead of noting individual cases of light-refraction, we can mentally reconstruct all present and future cases, if we know that the incident ray, the refracted ray, and the perpendicular lie in the same plane and that sin a/sin B = n. Here, instead of the numberless cases of refraction in different combinations of matter and under all different angles of incidence, we have simply to note the rule above stated and the values of n, which is much easier. The economical purpose is here unmistakable. In nature there is no law of refraction, only different cases of refraction. The law of refraction is a concise compendious rule, devised by us for the mental reconstruction of a fact, and only for its reconstruction in part, that is, on its geometrical side.

5. The sciences most highly developed economically are those whose facts are reducible to a few numberable elements of like nature. Such is the science of mechanics, in which we deal exclusively with spaces, times, and masses. The whole previously established economy of mathematics stands these sciences in stead. Mathematics may be defined as the economy of counting. Numbers are arrangement-signs which, for the sake of perspicuity and economy, are themselves arranged in a simple system. Numerical operations, it is found, are independent of the kind of objects operated on, and are consequently mastered once for all. When, for the first time, I have occasion to add five objects to seven others, I count the whole collection through, at once; but when I afterwards discover that I can start counting from 5, I save myself part of the trouble; and still later, remembering that 5 and 7 always count up to 12, I dispense with the numeration entirely. . . .

The mathematician who pursues his studies without clear views of this matter, must often have the uncomfortable feeling that his paper and pencil surpass him in intelligence. Mathematics, thus pursued as an object of instruction, is scarcely of more educational value than busying oneself with the Cabala. On the contrary, it induces a tendency toward mystery, which is pretty sure to bear its fruits.

6. The science of physics also furnishes examples of this economy of thought, although similar to those we have just examined. A brief reference here will suffice. The moment of inertia saves us the separate consideration of the individual particles of masses. By the force-function we dispense with the separate investigation of individual force-components. The simplicity of reasonings involving force functions springs from the fact that a great amount of mental work had to be performed before the discovery of the properties of the force-functions was possible. Gauss's dioptrics dispenses us from the separate consideration of the single refracting surfaces of a dioptrical system and substitutes for it the principal and modal points. But a careful consideration of the single surfaces had to precede the discovery of the principal and nodal points. Gauss's dioptrics simply saves us the necessity of often repeating this consideration.

We must admit, therefore, that there is no result of science which in point of principle could not have been arrived at wholly without methods. But, as a matter of fact, within the short span of a human life and with man's limited powers of memory, any stock of knowledge worthy of the name is unattainable except by the greatest mental economy. Science itself, therefore, may be regarded as a minimal problem, consisting of the most complete possible presentation of facts with the least possible expenditure of thought.

7. The function of science, as we take it, is to replace experience. Thus, on the one hand, science must remain in the province of experience, but, on the other, must hasten beyond it, constantly expecting confirmation, constantly expecting the reverse. Where neither confirmation nor refutation is possible, science is not concerned. Science acts and acts only in the domain of uncompleted experience. Exemplars of such branches of science are the theories of elasticity and of the conduction of heat, both of which ascribe to the smallest particles of matter only such properties as observation supplies in the study of the larger portions. The comparison of theory and experience may be farther and farther extended, as our means of observation increase in refinement.

Experience alone, without the ideas that are associated with it, would forever remain strange to us. Those ideas that hold good throughout the widest domains of research and that supplement the greatest amount of experience, are the most scientific. The principle of continuity, the use of which everywhere pervades modern inquiry, simply prescribes a mode of conception which conduces in the highest degree to the economy of thought.

8. If a long elastic rod be fastened in a vise, the rod may be made to execute slow vibrations. These are directly observable, can be seen, touched, and graphically recorded. If the rod be shortened, the vibrations will increase in rapidity and cannot be directly seen; the rod will present to the sight a blurred image. This is a new phenomenon. But the sensation of touch is still like that of the previous case; we can still make the rod record its movements; and if we mentally retain the conception of vibrations, we can still anticipate the results of experiments. On further shortening the rod the sensation of touch is altered; the rod begins to sound; again a new phenomenon is presented. But the phenomena do not all change at once; only this or that phenomenon changes; consequently the accompanying notion of vibration, which is not confined to any single one, is still serviceable, still economical. Even when the sound has reached so high a pitch and the vibrations have become so small that the previous means of observation are not of avail, we still advantageously imagine the sounding rod to perform vibrations, and can predict the vibrations of the dark lines in the spectrum of the polarized light of a rod of glass. If on the rod being further shortened all the phenomena suddenly passed into new phenomena, the conception of vibration would no longer be serviceable because it would no longer afford us a means of supplementing the new experiences by the previous ones.

When we mentally add to those actions of a human being which we can perceive, sensations and ideas like our own which we cannot perceive, the object of the idea we so form is economical. The idea makes experience intelligible to us; it supplements and supplants experience. This idea is not regarded as a great scientific discovery, only because its formation is so natural that every child conceives it. Now, this is exactly *what we do when we imagine a moving body which has just disappeared behind a pillar, or a comet at the moment invisible, as continuing its motion and retaining its previously observed properties. We do this that we may not be surprised by its reappearance. We fill out the gaps in experience by the ideas that experience suggests.*

9.   Yet not all the prevalent scientific theories originated so naturally and artlessly. Thus, chemical, electrical, and optical phenomena are explained by atoms. But the mental artifice atom was not formed by the principle of continuity; on the contrary, it is a product especially devised for the purpose in view. Atoms cannot be perceived by the senses; like all substances, they are things of thought. Furthermore, the atoms are invested with properties that absolutely contradict the attributes hitherto observed in bodies. However well fitted atomic theories may be to reproduce certain groups of facts, the physical inquirer who has laid to heart Newton's rules will only admit those theories as provisional helps, and will strive to attain, in some more natural way, a satisfactory substitute.

The atomic theory plays a part in physics similar to that of certain auxiliary concepts in mathematics; it is a mathematical model for facilitating the mental reproduction of facts. Although we represent vibrations by the harmonic formula, the phenomena of cooling by exponentials, falls by squares of times, etc., no one will fancy that vibrations in themselves have anything to do with the circular functions, or the motion of falling bodies with squares. It has simply been observed that the relations between the quantities investigated were similar to certain relations obtaining between familiar mathematical functions, and these more familiar ideas are employed as an easy means of supplementing experience. Natural phenomena whose relations are not similar to those of functions with which we are familiar, are at present very difficult to reconstruct. But the progress of mathematics may facilitate the matter.

As mathematical helps of this kind, spaces of more than three dimensions may be used, as I have elsewhere shown. But it is not necessary to regard these, on this account, as anything more than mental artifices.

This is the case, too, with all hypotheses formed for the explanation of new phenomena. Our conceptions of electricity fit in at once with the electrical phenomena, and take almost spontaneously the familiar course, the moment we note that things take place as if attracting and repelling fluids moved on the surface of the conductors. But these mental expedients have nothing whatever to do with the phenomenon itself.

As the preceding selection shows, many of Mach's fundamental assumptions are strikingly similar to Hume's.

First, for Mach, the raw material for all human knowledge of nature is our sensations: colors, tones, smells, etc. This commitment places him firmly in the Locke-Hume tradition of empiricism. The formula for avoiding both error and meaningless disputes about things we can never know is to trace our claims about nature to the sensations we have had.

Second, physical objects are simply clumps of sensations and are therefore our creation (since we do the "clumping") or, as Mach put it, "things of thought." Numbers are also simply human organizational instruments, abstracted originally from arrangements of sensations or clumps of sensations.

Third, causality, far from being a real relation of power between objects, is merely our shorthand for remembering that certain sensations or clumps of sensations have been found to occur before or after certain others in the past.

Fourth, the motive behind science is the economy of thought that makes prediction easier. Abstracting from our experience (that is, our immediate sensations) is useful because such generalization saves us time in anticipating what will come next.

If we are to be faithful to our experience, are we to abandon scientific imagination and speculative theorizing? Mach sees no necessity to do so if we confine our imagination to what is suggested by experience. Laws are simply summaries of observations from the past coupled with plausible extensions of these observations to cases we haven't been able to observe. Here, again, economical generalization is the means to better prediction. Notice in Mach's own illustrations his own caution about the use of theories, models, and the scientific imagination in general (for instance, his different reactions to the imagination's role concerning vibration on the one hand and atoms on the other).

An adequate explanation is, then, according to Mach, a generalized description and an improvement over the description of the individual sensations themselves because it saves time. To explain is to remind us of our past experiences by describing them and thereby allowing us to anticipate, or predict, future experience.

The core of Mach's position is this: our theories, our imaginative models do not by themselves have any claim to truth, to a correspondence with reality. Our thoughts derive whatever truth they have only by their representation of our experience. We do not receive or develop an intellectual intuition, apart from experience, that reveals reality to us. Thinking is not another way of seeing. It serves best merely as a way to generalize what we have actually seen with our eyes and apprehended by our other senses. Yet for some positivists this demand that theorizing has its validity only in its predictiveness allowed some new liberty for theorists. If a theory need not be true on its own grounds, if it need not correspond to reality except by predicting, then one could allow any sort of theory, and any sort of entities that one would care to invent, so long as they held hope of being justified as predictive and economical. Flights of imagination might be allowed, as long as we remembered not to take them seriously apart from their predictive success.

Pierre Duhem summarized the positivist's view of theories and their relation to experience as follows:

> Thus a true theory is not a theory which gives an explanation of a physical appearance in conformity to reality; it is a theory which represents in a satisfactory manner a group of experimental laws. A false theory is not an attempt at an explanation based on assumptions contrary to reality; it is a group of propositions which do not agree with the experimental laws. *Agreement with* experiment is the sole criterion of truth for a physical theory.[3]

One further implication of the positivist view should be made explicit, namely, that explanation, description, and prediction are not merely related in meaning. Rather, explanation and prediction are both defined in terms of description. Explanation is generalized description, description of a whole class of events, and prediction is simply description of a future event. We sometimes are tempted to think of an explanation as *leading to* predictions. Notice, however, that in Mach's view an explanation will consist of predictions. It will be a description of a class of events some of which have not yet occurred. Predictions are not, then, the *tests* of an explanation's adequacy so much as a part of the explanation itself. To say that water boils at 100 degrees Celsius is a summary of past experience and a prediction concerning future events. It is explained by kinetic molecular theory only insofar as that theory provides more general descriptions of a class of phenomena among which is included boiling water. The positivist's point of view recommends, then, that we modify our use of some of these key terms in science. The positivist suggests that whatever meaning is assigned to the term *explanation* beyond the meaning "generalized description" is going to confuse us, tempt us to go beyond what we can justify, and lead us to pervert and misunderstand the proper methods of science.

Phenomenalism: The Positivist Response to Skepticism

Hume's Skepticism.   David Hume believed that his arguments (see Chapter 3) lead us to skepticism. He did not recommend that we give up thinking, but he did believe that our thinking turns out to be a "species of feeling" rather than a source of knowledge. Why did he believe this? Let us use the following diagram to clarify the position that Hume was attacking and to introduce some helpful new terms. When we experience something—for example, that it is snowing outside—we might chart what happens this way:

We – – – – have – – – – perceptions – – – – of – – – snow

(our minds)   (whiteness and   (the real
                       downward motions of   thing behind
                       whiteness, etc.)   our perceptions)

or, more generally,

Mind – – – – has – – – – perceptions – – – – of – – – – reality.

The most persuasive arguments for the accuracy of this diagram are that (1) we would normally take the snow to be a reality that is independent of perceptions of it; i.e., it would in fact be snowing even if we weren't perceiving it; and (2) what we actually *see*, or perceive, are such things as specks of whiteness, downward motion, etc., which are qualities that we attribute to the snow but are, after all, simply sensations that we are having.

Now, what are we to answer when someone asks what the snow is like in itself and apart from our perceptions of it? We will no longer be able to say, "Well, it's white and moving downward," since these are references to our perceptions. The questioner wants to know what snow is like *independent* of such perceptions. Notice that this question might be put in terms of the correspondence and coherence theories of truth discussed in Chapter 3. We might ask how we know that our perceptions *correspond* to reality.

The traditional answer of Hume's opponents, the rationalists, was that we know of the existence of such realities by our reason, not by means of our changeable senses. There was wide agreement among rationalists and empiricists that such sensations as redness were subjective and not really in the object itself, but many rationalists thought that some qualities (for instance such universal ones as taking up space and moving in space) were real properties of real independent objects and that this was knowable by reason.

Hume was not interested in arguing about what sort of entities were "really out there." Such arguments are called *ontological* because they refer to "being, as it is in itself" (from the Greek *ontos*, or "being"). He thought that such arguments were useless because neither reason nor the senses could gain access to such realities. If all we are familiar with are our own perceptions, then no number of these perceptions can tell us what lies *behind* them. Reason, being simply a manipulator of perception, can be of no help here either.

Hume's argument concerns *epistemology*, or the theory of knowledge. He answered the question "what can we know?" by saying, "only our perceptions." But such an answer seems to lead us to say that no ontological knowledge is possible at all. We might say that you can never get behind a perception of whiteness to any knowledge of the snow itself.[4] Similarly, we might conclude that we can never get behind our perceptions, or experiences, to any knowledge that lies behind them whatever. We saw one illustration of this skeptical conclusion in Hume's discussion of causality. However many repetitions of a sequence of perceptions we may have, these never justify ontological claims about causes in nature, according to Hume.

Positivism and Skepticism. We have portrayed positivism as concerned with descriptive scientific knowledge. Mach was concerned with liberating science from metaphysical speculations. Whereas *metaphysics* originally meant the study of being,[5] it became for Mach and the other positivists a term of rebuke for speculations having no ground in experience. He wanted to liberate science not in order to destroy the possibility of knowledge but to allow science to get on with the job of acquiring knowledge. Yet here we encounter a central problem. *Did not Mach's general agreement with Hume force him to the same skeptical conclusions as Hume felt forced to adopt?* If, as Mach said, the foundation of science is sensation, how can we ever develop ontological knowledge, that is, knowledge of the realities that account for our having certain sensations? The position that Hume and Mach adopted concerning what we can know seems to have denied the possibility of our having any access to the real world.

The Doctrine of Phenomenalism. Mach and many other positivists were far from being ready to accept Hume's skepticism. They found a way to avoid such a conclusion by redrawing the diagram we considered earlier, following the theory of British philosopher George Berkeley.

Berkeley's theory, called phenomenalism, agrees that all we are aware of is perceptions. But Berkeley argued that we don't need to make ontological claims concerning the alleged source of these perceptions.[6] Suppose that instead of seeing the relationships involved in knowing as

Mind – – – – has – – – – perceptions – – – – of – – – – reality,

we redo our image of knowing as follows:

Mind – – – – has – – – – perceptions

Reality

In this view, our perceptions are no longer alleged to *represent* reality or to be produced by a reality behind them. Rather, they, with the mind that has them, *are* reality. The strength of phenomenalism is that it interprets the world as knowable because we do have direct access to our perceptions. No longer is there a reality that we merely hope is represented by our perceptions but that we can't observe directly. What is the chair in which you are sitting really like? Well, it's brown, perhaps, hard, composed of wood. Oh, but those are only your perceptions of it. In a different light it would no longer be brown. To an ant, it is no doubt less smooth than it appears to you. Apart, then, from how it looks to you, what is it like? Berkely suggested that this line of questions, which seems reasonable, isn't. It's nonsense. The chair is a collection of sensations, labeled for convenience and economy, and that is all it is. To exist *means* either to perceive or to be perceived.

To those who object that we have not, on this account, located the cause of perceptions, Mach would reply that the notion of cause is merely a human abstraction from perceived sequences. Only the earlier incorrect diagram deludes us into believing that the further search for causes behind our perceptions makes sense or is worthwhile.

Relativism as the Threat to Phenomenalism

The phenomenalist doctrine, that our perceptions are reality, attempts to abolish the distinction we have made between ontological questions and epistemological questions. The questions "what can we know" and "what is real" have the same answer: our perceptions. If such a view has no serious faults, it provides a strong alternative to skepticism for the positivist. Unfortunately, one major snag must be dealt with before phenomenalism can be of much use to the positivist. We will remember that Mach and other positivists fear the intrusions of metaphysical speculation in scientific inquiry. Why? One major reason is that the positivists believe that metaphysical arguments aren't publicly decidable. We all seem to have our own metaphysical opinions, with no method of agreement with everyone else. This leads to relativism—the situation in which the truth seems to be relative to each person's assumptions, with no way of moving toward consensus. But, ironically, the phenomenalist doctrine, which might help positivists avoid skepticism, seems itself highly vulnerable to becoming relativistic. While I need not be skeptical concerning my own perceptions—it makes no sense to say that I am wrong in seeing the salmon-colored sky there in the west—it is still quite possible that you do not see the same salmon color.

And if we disagree, how in the world can the dispute be resolved? In fact, would any resolution of the dispute make sense? Perhaps you will accept the referee's claim that you aren't really seeing the shade of pink you think you are seeing, but I am *quite* sure that I am seeing just that salmon shade I told you about!

How can we interpret science, which demands repeatability of experiments and at least much broad agreement among participants, as based on a standard as fleeting and subjective as individual sense perceptions? Such a view seems both false, as a description of how science reaches agreement, and dangerous, if taken as how science ought to reach agreement. Phenomenalists have attempted to explain the universality and uniformity of phenomena (and thereby circumvent the threat of relativism) in a variety of ways without resorting to the alleged myth of the "real" objects behind the phenomena. Berkeley himself had recourse to a conception of God as the rational perceiver, who by willing an orderly set of perceptions makes it possible for us all to see the same order. We saw in Chapter 3 that Kant felt that the structure of our minds provides invariance and intersubjectivity concerning some of the most necessary kinds of perception.[7]

The positivist who embraces phenomenalism must solve two distinct problems. Both relate to the question of relativism. First, there is the question of the *uniformity* of our perceptions or how we can be sure that nature is uniform. Can we know that our perceptions will repeat themselves in reliable patterns in the future? If not, how can we ever be justified in making predictions based on past experience? Second, we must assure ourselves of the *universality* of our perceptions, that is, on our ability to agree with one another on what we are perceiving, even though sensations seem to vary considerably from person to person.

The Problem of Uniformity. In trying to account for the uniformity of nature, one comes again to appreciate the importance of the causal principle attacked by Hume. A knowledge of causes in nature might provide knowledge of ongoing uniform powers to produce natural effects. That could give us confidence that the future will, in general outline, be like the past. But phenomenalists, following Hume's critique, do not believe that we have such knowledge. How then can the uniformity of nature be accepted? We have seen the range of responses to this question implicitly in the responses to Hume considered in Chapter 3. One may hold, as Hume did, that belief in the uniformity of nature is a feeling rather than a rationally held conclusion. In this view, we need not seek reasons for believing in uniformity since we believe in it regardless of reason. We don't need to justify the belief any more than we need to justify liking the taste of chocolate ice cream. Both are ultimately matters of natural feeling. While this approach seems to eliminate the need for a justification of belief in uniformity, it places such a large stone of nonrationality in the foundation of scientific explanation that few scientists would remain comfortable with it.

The Kantian alternative is to view the uniformity of nature (i.e., the predictability of persistent kinds of perception) as a reflection of the ordering structure of the mind. The world, for us, will always be causal, Kant said, because our mind can think of our perceptions, literally, in no other way. While this approach distinguishes thinking and feeling rather than merging them as Hume did, its plausibility depends on the adequacy of the Kantian argument that there really are such structures of the mind.

Kant's theory is attractive as a starting point to those who want to avoid both the subjectivism of Hume's "belief as a feeling" theory and, at the other pole, the affirmation of knowledge of the reality behind and independent of human perception.

Mach, in section 3 of the selection quoted earlier in this chapter, agreed with Hume and Kant on the human propensity for affirming uniformity in causal relations without explicitly committing himself to either alternative they represent:

> We may, indeed, say that our sense of causality is not acquired by the
> individual, but has been perfected in the development of the race. Cause
> and effect, therefore, are things of thought having an economical office. It
> cannot be said *why* they arise. For it is precisely by the abstraction of
> uniformities that we know the question "why."

Perhaps this is as far as the practicing scientist need take the inquiry concerning uniformity. One might say, "I'll seek uniformities in nature (whether nature is my perceptions or the "realities" behind those perceptions), accepting such inquiry as the normal propensity of the inquiring scientific mind. I'll leave it to philosophers of science to work out the justification for such a set of assumptions."

In fact, identifying the use of the causal principle with what we have here referred to as the assumption of the uniformity of nature, Ernest Nagel believes that practicing scientists do go ahead and make the assumption, justified or not:

> Nevertheless, the actual pursuit of theoretical science in modern times is
> directed towards certain goals, one of which is formulated by the principle
> of causality . . . it is difficult to understand how it would be possible for
> modern theoretical science to surrender the general ideal expressed by the
> principle without becoming thereby transformed into something incomparably different from what that enterprise actually is.[8]

The problem of uniformity seems to present a cruel dilemma to those who are attempting to make sense of scientific explanation. If we believe with Mach that whatever knowledge we can have of nature comes only from experience, we appear at this point unable to provide any justification whatever of causes in nature or any other features of nature that will give us confidence in continuing uniformity of any kind. If, on the other hand, we continue to seek justification for expecting uniformities in nature to continue, we will need to do so by defending our knowledge of nature against Hume's skepticism. And this may necessitate abandoning an exclusive reliance on experience as the source of knowledge.

The Problem of Universality and the Doctrine of the Given. Sense experience seemed to the positivists best suited to provide the dispassionate, objective basis for a science liberated from subjective metaphysical speculations. The ideal is direct responsibility to the facts of sense experience. These sense data *must* then be the same for all of us and unaffected themselves by theory. Mach concedes that in our investigation of experience we select what we are most interested in and thereby affect the results. However, he insists that in good scientific practice our hypotheses must not be seen as affecting the reality of our experience. It is the facts that must determine scientific truth.

Does it make sense to say that there is such a universal (that is, sufficiently intersubjective) nontheoretical "given" in our sense experience? Are there irreducible, primary, simple, and indisputable facts? Critics of positivism say that there are not and that when positivists assume that we can simply rely on the facts on this basic level they are blinded to the real interplay between theories and facts and to the importance of theories in general. Put simply, the matter stands as follows. The positivists must claim that some sensations are given to all of us similarly in order to provide for the broad agreement science needs. This means that we need cold, hard facts. But if no such universal givens are available, if our view of the facts is always assumption bound or theory involved, then it seems futile to appeal to "the facts" or "sensations" alone to account for the universality that science needs. And if positivism can't account for any universality in scientific explanation, it will be doomed to the relativism it feared.

How can we decide whether or not there are given indisputable facts—unarguable observations—concerning our sensations that could resolve scientific disputes? It has been suggested that if such givens exist, then we ought to be able to decide between any two competing theories by appeal to the facts. That is, if theories are tested exclusively by whether they represent experimental laws and if these laws are only summaries of past and future experience, then experiments revealing those experiences should allow us to decide between the theories. For this reason, such experiments, if they exist, can be called crucial experiments.

In the selection that follows, Irving Copi examines the question of crucial experiments.[9]

It might appear that, given any problem, all one needs to do is set down all relevant hypotheses and then perform a series of crucial experiments to eliminate all but one of them. The surviving hypothesis is then "the answer," and we are ready to go on to the next problem. But no opinion could possibly be more mistaken.

It has already been remarked that formulating or discovering relevant hypotheses is not a mechanical process but a creative one: some hypotheses require genius for their discovery. It has been observed further that crucial experiments may not always be possible, either because no different observable consequences are deducible from the alternative hypotheses, or because we lack the power to arrange the experimental circumstances in which different consequences would manifest themselves. We wish at this time to point out a more pervasive theoretical difficulty with the program of deciding between rival hypotheses by means of crucial experiments. It may be well to illustrate our discussion by means of a fairly simple example. One that is familiar to all of us concerns the shape of the earth.

In ancient Greece, the philosophers Anaximenes and Empedocles had held that the earth is flat, and this view, close to common sense, still had adherents in the Middle Ages and the Renaissance. Christopher Columbus, however, insisted that the earth is round—or rather, spherical.

One of Columbus' arguments was that as a ship sails away from shore, the upper portions of it remain visible to a watcher on land long after its lower parts have disappeared from view. A slightly different version of the same argument was included by Nikolaus Copernicus in his epoch-making treatise On the Revolutions of the Heavenly Spheres. In Section II of Book I of that work, entitled "That the Earth Also Is Spherical," he presented a number of arguments intended to establish the truth of that view. Of the many found there we quote the following:

> That the seas take a spherical form is perceived by navigators. For when land is still not discernible from a vessel's deck, it is from the masthead. And if, when a ship sails from land, a torch be fastened from the masthead, it appears to watchers on the land to go downward little by little until it entirely disappears, like a heavenly body setting.

As between these two rival hypotheses about the earth's shape, we might regard the foregoing as a description of a crucial experiment. The general pattern is clear. From the hypothesis that the earth is flat, H_f, it follows that if a ship gradually recedes from view, then neither its masthead nor its decks should remain visible after the other has vanished. On the other hand, from the hypothesis that the earth is spherical, H_s, it follows that if a ship gradually recedes from view, its masthead should remain visible after its decks have vanished from sight. The rationale involved here is nicely represented by a diagram.

a   b

In the figure, a represents the situation which would obtain if H_f were true. It is clear that if the earth is flat there is no reason why any one portion of the ship should disappear from sight before any other portion. The figure b represents the situation corresponding to H_s. As the ship recedes, the curvature of the earth rises between the observer and the ship, blocking out his view of the decks while the masthead still remains visible. In each case the rays of light passing from ship to observer are represented by dotted lines. Now the experiment is performed, a receding ship is watched attentively, and the masthead does remain visible after the decks have disappeared. Our experiment may not have demonstrated the truth of H_s, it can be admitted, but surely it has established the falsehood of H_f. We have as clear an example of a crucial experiment as it is possible to obtain.

But the experiment described is not crucial. It is entirely possible to accept the observed facts and still maintain that the earth is flat. The experiment has considerable value as evidence, but it is not decisive. It is not crucial because the various testable predictions were not inferred from the stated hypotheses H_f and H_s alone, but from them plus the additional hypothesis that light travels in straight lines. The diagrams show clearly that this additional assumption is essential to the argument. That the decks disappear before the masthead does is not deducible from H_s alone but requires the additional premise that light rays follow a rectilinear path (H_r). And that the decks do not disappear before the masthead does is not deducible from H_f alone but requires the same additional premise: that light rays follow a rectilinear path (H_r). The latter argument may be formulated as:

The earth is flat (H_f).
Light rays follow a rectilinear path (H_r).
Therefore the decks of a receding ship will not disappear from view before the masthead.

Here is a perfectly good argument whose conclusion is observed to be false. Its premises can not both be true; at least one of them must be false. But which one? We can maintain the truth of the first premise, H_f, if we are willing to reject the second premise, H_r. The second premise, after all, is not a truth of logic but a contingent statement which is easily conceived to be false. If we adopt the contrary hypothesis that light rays follow a curved path, concave upwards, (H_c), what follows as conclusion now? Here we can infer the denial of the conclusion of the former argument. From H_f and H_c it follows that the decks of a receding ship will disappear before its masthead does. The following figure explains the reasoning involved here.

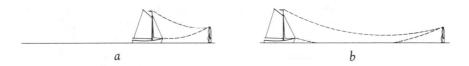

a   b

In this figure a represents the situation when the ship is near the shore, while b shows that as the ship recedes, the earth (even though flat) blocks out the view of the decks while the masthead still remains visible. The light rays in this diagram too are represented by dotted lines, but in this case curved rather than rectilinear. The same experiment is performed, the decks do disappear before the masthead, and the observed fact is perfectly compatible with this group of hypotheses which includes H_f, the claim that the earth is flat. The experiment, therefore, is not crucial with respect to H_f, for that hypothesis can be maintained as true regardless of the experiment's outcome. (This illustration was first suggested to me by my colleague Professor C. L. Stevenson.)

The point is that where hypotheses of a fairly high level of abstractness or generality are involved, no observable or directly testable prediction can be deduced from just a single one of them. A whole group of hypotheses must be used as premises, and if the observed facts are other than those predicted, at least one of the hypotheses in the group is false. But we have not established which one is in error. An experiment can be crucial in showing the untenability of a group of hypotheses. But such a group will usually contain a considerable number of separate hypotheses, the truth of any one of which can be maintained in the teeth of any experimental result, however "unfavorable," by the simple expedient of rejecting some other hypothesis of the group. A conclusion often drawn from these considerations is that no individual hypothesis can ever be subjected to a crucial experiment.

. . . even if we confine our attention to theoretically significant hypotheses, and never invoke any ad hoc hypotheses at all, no experiments are ever crucial for individual hypotheses, since hypotheses are testable only in groups. Our limitation here serves to illuminate again the systematic character of science. Scientific progress consists in building ever more adequate theories to account for the facts of experience. True enough, it is of value to collect or verify isolated particular facts, for the ultimate basis of science is factual. But the theoretical structure of science grows in a more organic fashion. In the realm of theory, piecemeal progress, one step at a time advances, can be accomplished, but only within the framework of a generally accepted body of scientific theory. The notion that scientific hypotheses, theories, or laws are wholly discrete and independent is a naive and outdated view.

The term "crucial experiment" is not a useless one, however. Within the framework of accepted scientific theory which we are not concerned to question, a hypothesis can be subjected to a crucial experiment. If a negative result is obtained, that is, if some phenomenon fails to occur which had been predicted on the basis of the single dubious hypothesis together with accepted parts of scientific theory, then the experiment is crucial and the hypothesis is rejected. But there is nothing absolute about such a procedure, for even well-accepted scientific theories tend to be changed in the face of new and contrary evidence. Science is not monolithic, either in its practices or in its aims.

Perhaps the most significant lesson to be learned from the preceding discussion is the importance to scientific progress of dragging "hidden assumptions" into the open. That light travels in straight lines was assumed in the arguments of Columbus and Copernicus, but it was a hidden assumption. Because they are hidden, there is no chance to examine such assumptions critically and to decide intelligently whether they are true or false. Progress is often achieved by formulating explicitly an assumption which had previously been hidden and then scrutinizing and rejecting it. An important and dramatic instance of this occurred when Einstein challenged the universally accepted assumption that it always makes sense to say of two events that they occurred at the same time. In considering how an observer could discover whether or not two distant events occurred "at the same time," Einstein was led to the conclusion that two events could be simultaneous for some observers but not for others, depending upon their locations and velocities relative to the events in question. Rejecting the assumption led to the Special Theory of Relativity, which constituted a tremendous step forward in explaining such phenomena as those revealed by the Michelson-Morley experiment. It *is clear that an assumption must be recognized before it can be challenged. Hence it is enormously important in science to formulate explicitly all relevant assumptions in any hypothesis, allowing none of them to remain hidden.*

If Copi's argument is accepted, then the ideal of a purely observational, nontheoretical set of explanations will have to be abandoned. Nature is describable, if we agree with him, only within the context of a set of commitments that we might call, following popular usage, a conceptual framework. This means that agreement about which scientific explanations to accept will depend in part upon what assumptions and theories we accept. But this seems to lead us in a different direction from Mach, who claims that except for economy, theories are dispensable in favor of pure observation.

Ernest Nagel, a contemporary philosopher of science quoted earlier, summarizes the criticism of the positivistic appeal to the facts alone:

> As a matter of psychological fact, elementary sense data are not the primitive materials of experience out of which all our ideas are built like houses out of initially isolated bricks. On the contrary, sense experience normally is a response to complex though unanalyzed patterns of qualities and relations; and the response usually involves exercise of habits of interpretation and recognition based on tacit beliefs and inferences, which cannot be warranted by any single momentary experience. Accordingly, the language we normally use to describe even our immediate experiences is the common language of social communication, embodying distinctions and assumptions grounded in a large and collective experience, and not a language whose meaning is supposedly fixed by reference to conceptually uninterpreted atoms of sensation.
>
> It is indeed sometimes possible under carefully controlled conditions to identify simple qualities that are directly apprehended through the sense organs. But the identification is the terminus of a deliberate and often difficult process of isolation and abstraction, undertaken for analytical purposes; and there is no good evidence to show that sensory qualities are apprehended as atomic symbols except as the outcome of such a process. Moreover, though we may baptize these products by calling them sense data and may assign different labels to different classes of them, the use and meanings of those names cannot be established except by way of directions for instituting processes involving overt bodily activities. Accordingly, the meanings of sense data terms can be understood only if the distinctions and assumptions of our commerce with the gross objects of experience are taken for granted. In effect, therefore, those terms can be used and applied only as part of the vocabulary of common sense. In short, the "language" of sense data is not an autonomous language, and no one has yet succeeded in constructing such a language.[10]

From the above, it is clear that some philosophers of science are going to quarrel with the positivist thesis that explanation is simply general description. If describing the facts cannot be done without theoretical commitments, then perhaps explanation schemes, which these theories provide, are prior to and different from description in important ways. We may want to continue to insist that explanation involves description but also seek to locate what else it involves. For instance, explanation schemes or conceptual frameworks may include a logical structure in addition to a Machian simple description. These further questions about the logical structure of scientific explanation, and what they include that cannot be called mere description, can interest both the theorist who rejects positivism as incorrect and the positivist who is apprehensive about and therefore interested in the added features that distinguish explanation from a description of sensory experiences. The nonpositivists will be looking for new sources of knowledge apart from summaries and observations. The positivists will be seeking to show that whatever explanation includes beyond description concerns only the structure, not the content, of knowledge. They will continue to insist that the content is derived exclusively from observations.

The possibility that the logical structure of competing scientific explanations is of crucial importance in choosing the best one has led to extensive discussion among philosophers of science. In recent years one particular view of the structure of explanations has been the subject of much debate. In Chapter 5 we will consider this proposal, which in some ways can be considered an enrichment and in other ways perhaps an alternative to the concepts of explanation that we have already considered.

Supplementary Readings

Edwards, Paul, ed. *The Encyclopedia of Philosophy.* New York: Collier Macmillan Publishers, 1967.

The entry on Mach (vol. 5, pp. 115–119) by Peter Alexander is an example of what can be found in this important work. It offers a good description of Mach as a scientist and a philosopher, provides a clear statement of his views, and with most entries, includes a bibliography.

von Mises, Richard. *Positivism: A Study in Human Understanding.* New York: George Braziller, 1956.

This book is an excellent statement of mature positivism. Its introduction gives a clear account of the argument of the whole book. Each of the numbered sections of the book is summarized with clarity.

Notes

1. Auguste Comte, *The Essential Comte: Selections from Cours de Philosophie Positive,* ed. S. Andreski, trans. M. Clarke (New York: Barnes & Noble, 1974).

2. Ernst Mach, *The Science of Mechanics,* 6th English ed., trans. T. J. McCormack (LaSalle, Ill.: Open Court Publishing Co., 1960), pp. 577–590. Footnotes have been omitted.

3. Pierre Duhem, *The Aim and Structure of Physical Theory*, 2d ed., 1914, trans. Philip Wiener (Princeton, N.J.: Princeton University Press, 1954), pp. 20–21. This work, like Mach's, is a landmark in the literature of the philosophy of science. Duhem presents with clarity and persuasiveness the hope, shared by positivists, that scientific theories may find universal acceptance or rejection by being liberated from dependence on any particular ontological position. (In many respects, however, Duhem was not a positivist. His position is a complex and powerful one that deserves attention in its own right and certainly resists easy classification.)

4. We may be tempted to answer this puzzle by saying, "We know that the snow is frozen water reflecting light, which accounts for the perception of whiteness, and therefore is a reality that stands behind our perceptions." Hume would remind us, however, that such a term as *frozen* is itself defined simply in terms of collections of perceptions. We would therefore have shifted the discussion to *other* perceptions but certainly not gotten behind perceptions in general.

5. *Metaphysics* was the name an early editor, Andronicus of Rhodes, gave to Aristotle's work on "first philosophy." See Richard McKeon, *The Basic Works of Aristotle* (New York: Random House, 1941), p. xviii.

6. For a systematic defense of this position, consult either of two works of George Berkeley, *A Treatise Concerning the Principles of Human Knowledge*, 1710 (Indianapolis: Bobbs-Merrill, 1957); or *Three Dialogues Between Hylas and Philonous*, 1713 (Indianapolis: Bobbs-Merrill, 1954).

7. For a number of reasons, Kant was not a phenomenalist. For instance, he believed that for certain purposes we need to postulate the existence of realities beyond experience. Nevertheless, he was indebted to Berkeley for many ideas concerning the nature of our phenomenal experience.

8. Ernest Nagel, *The Structure of Science* (New York: Harcourt, Brace and World, 1961), p. 324.

9. Irving Copi, *Introduction to Logic*, 3d ed. (New York: Macmillan, 1968), pp. 400–406; footnotes omitted.

10. Nagel, *The Structure of Science*, pp. 121–122.

/5/ THE COVERING LAW MODEL

Proposal: The Most Adequate Scientific Explanation Consists in the Deduction of What Is to Be Explained from Covering Laws

We began by seeing scientific explanation as the locating of causes (Chapter 3). We wanted to find the reliable sources of change in nature. Yet Hume argued that such a search was in vain, and none of the responses to his argument seemed likely to win overwhelming approval. Then a radical solution presented itself. Perhaps all such inquiry into causes is doomed from the start because we can only be sure of our sensations, not of the rational structure behind those sensations. This possibility led us to consider the positivist program (Chapter 4), in which explanation is identified simply as generalized description. In order to avoid the skeptical conclusions of Hume, however, positivists such as Mach embraced the phenomenalist doctrine that our perceptions are all there is to reality; and phenomenalism seemed likely to lead to the sort of relativism that cannot account for either the uniformity of our perceptions through time, or universality from one person to another. Accounting for uniformity meant either making assumptions that we did not even try to justify or making assertions that would need extensive further defense. And accounting for universality seemed likely to commit us to a belief in uninterpreted, indisputable observations, which critics warned us might not be justifiable.

For many theorists, these problems, which we have seen both in causal accounts of explanation and in positivism, are just that—problems, not fatal objections. They may be seen as reasons to refine or modify causal or positivistic theories or as objections that may be overcome by more careful future analysis. No theory is likely to be rejected only because it has flaws or unresolved problems. However, other thinkers have tried very different approaches to the quest for an understanding of scientific explanation, and the objections to the theories we have discussed are, for some, an important reason for trying a different tack.

So far, we've discussed theories that locate explanations by reference to what they are about (causes or phenomena), in short, their *content*. A promising and popular alternative, particularly in the last several decades, has been to focus on their *logical structure*. One reason for looking in this direction was suggested at the conclusion of Chapter 4; if our observations always depend in part on a prior acceptance of many assumptions and these assumptions are imbedded in networks that we call theories, then an explanation of observations will have to depend on the reliability of these networks. An important feature of the networks is how their elements are logically related to each other and how the elements are logically related to experimental laws and observations. It is therefore likely that the logical structure of a theory network will tell us something about its adequacy.

The logical structure of explanations also seems important when we consider what we would like explanations to do for us. When some phenomenon is explained, we usually expect to learn something that we didn't know before about the connection between the phenomenon and something else, perhaps laws or other phenomena. For example, suppose you want an explanation for the appearance of a rainbow. Presumably, to say that it is a multicolored arc in the sky would not explain it. After all, in that sense you already know what it is. You might expect to learn what *kind* of thing it is and enough about how that kind of entity behaves so that you might be *able to predict future occurrences* of it, or at least the likelihood of future occurrences. You might expect (reminiscent of our discussion of causality) to learn what makes a rainbow occur. Now what sort of information would help? Suppose you were told that a rainbow is a refraction spectrum, resulting from light rays from the sun being reflected and separated by the droplets of water in the atmosphere. This explanation apparently gives you three pieces of information: (1) it tells you what the rainbow is (a refractive phenomenon); (2) it identifies conditions that must exist for a rainbow to occur (water droplets, light rays); and (3) it seems implicitly to call on a law (droplets influence light in certain general ways). Moreover, all of these elements seem necessary if your hopes are to be fulfilled. Without identification of the rainbow as a refractive phenomenon, you don't know which law to refer to. Without the antecedent conditions, the law finds no application. Without the law, you can neither understand the rainbow as an instance of a general pattern nor have any reason to expect another one in the future.

What logical relationship exists between the statement that the rainbow is observed and the other elements in the explanation? That the relationship is *deductive* seems clear.[1] If one has the appropriate antecedent conditions (water droplets and light rays in certain specifiable relationships to each other), then the law "guarantees" the occurrence of the rainbow.

The adequacy of the foregoing explanation (in answering the questions we ask of an explanation) suggests that (1) the logical character of an explanation is the key to its adequacy and (2) the crucial feature of this logical character is a deductive relationship between the laws and antecedent conditions, as premises,

and the phenomenon to be explained, as a conclusion. This theory of explanation is called the deductive-nomological model (*nomological* means "referring to
laws"), or the *covering law model*.[2]

Although the covering law model has been proposed by many, including
John Stuart Mill in his *System of Logic* (1843), its most succinct exposition is
contained in a widely read essay by Carl Hempel and Paul Oppenheim. We have
included here only the first part of the essay as it was originally published.[3]

STUDIES IN THE LOGIC OF EXPLANATION

1.   Introduction. *To explain the phenomena in the world of our experience, to answer the question "why?" rather than only the question "what?" is one of the foremost
objectives of all rational inquiry; and especially, scientific research in its various
branches strives to go beyond a mere description of its subject matter by providing an
explanation of the phenomena it investigates. While there is rather general agreement
about this chief objective of science, there exists considerable difference of opinion as to
the function and the essential characteristics of scientific explanation. In the present
essay, an attempt will be made to shed some light on these issues by means of an
elementary survey of the basic pattern of scientific explanation and a subsequent more
rigorous analysis of the concept of law and of the logical structure of explanatory
arguments. . . .*

ELEMENTARY SURVEY OF SCIENTIFIC EXPLANATION

2.   Some illustrations. *A mercury thermometer is rapidly immersed in hot water;
there occurs a temporary drop of the mercury column, which is then followed by a
swift rise. How is this phenomenon to be explained? The increase in temperature
affects at first only the glass tube of the thermometer; it expands and thus provides a
larger space for the mercury inside, whose surface therefore drops. As soon as by heat
conduction the rise in temperature reaches the mercury, however, the latter expands,
and as its coefficient of expansion is considerably larger than that of glass, a rise of the
mercury level results. This account consists of statements of two kinds. Those of the
first kind indicate certain conditions which are realized prior to, or at the same time as,
the phenomenon to be explained; we shall refer to them briefly as antecedent conditions. In our illustration, the antecedent conditions include, among others, the fact that
the thermometer consists of a glass tube which is partly filled with mercury, and that it
is immersed into hot water. The statements of the second kind express certain general
laws; in our case, these include the laws of the thermic expansion of mercury and of
glass, and a statement about the small thermic conductivity of glass. The two sets of
statements, if adequately and completely formulated, explain the phenomenon under
consideration: they entail the consequences that the mercury will first drop, then rise.
Thus, the event under discussion is explained by subsuming it under general laws, i.e.,
by showing that it occurred in accordance with those laws, by virtue of the realization
of certain specified antecedent conditions.*

*Consider another illustration. To an observer in a row boat, that part of an oar
which is under water appears to be bent upwards. The phenomenon is explained by*

means of general laws—mainly the law of refraction and the law that water is an optically denser medium than air—and by reference to certain antecedent conditions—especially the facts that part of the oar is in the water, part in the air, and that the oar is practically a straight piece of wood. Thus, here again, the question "Why does the phenomenon happen?" is construed as meaning "according to what general laws, and by virtue of what antecedent conditions does this phenomenon occur?"

So far, we have considered exclusively the explanation of particular events occurring at a certain time and place. But the question "Why?" may be raised also in regard to general laws. Thus, in our last illustration, the question might be asked: Why does the propagation of light conform to the law of refraction? Classical physics answers in terms of the undulatory theory of light, i.e., by stating that the propagation of light is a wave phenomenon of a certain general type, and that all wave phenomena of that type satisfy the law of refraction. Thus, the explanation of a general regularity consists in subsuming it under another, more comprehensive regularity, under a more general law. Similarly, the validity of Galileo's law for the free fall of bodies near the earth's surface can be explained by deducing it from a more comprehensive set of laws, namely, Newton's laws of motion and his law of gravitation, together with some statements about particular facts, namely, the mass and the radius of the earth.

3. The Basic Pattern of Scientific Explanation. From the preceding sample cases let us now abstract some general characteristics of scientific explanation. We divide an explanation into two major constituents, the explanandum and the explanans. By the explanandum, we understand the sentence describing the phenomenon to be explained (not that phenomenon itself); by the explanans, the class of those sentences which are adduced to account for the phenomenon. As was noted before, the explanans falls into two subclasses; one of these contains certain sentences $C_1, C_2, \ldots, C_k$ which state specific antecedent conditions; the other is a set of sentences $L_1, L_2, \ldots, L_r$ which represent general laws.

If a proposed explanation is to be sound, its constituents have to satisfy certain conditions of adequacy, which may be divided into logical and empirical conditions. For the following discussion, it will be sufficient to formulate these requirements in a slightly vague manner; in Part III, a more rigorous analysis and a more precise restatement of these criteria will be presented.

I. Logical Conditions of Adequacy

(R1) The explanandum must be a logical consequence of the explanans; in other words, the explanandum must be logically deducible from the information contained in the explanans, for otherwise, the explanans would not constitute adequate grounds for the explanandum.

(R2) The explanans must contain general laws, and these must actually be required for the derivation of the explanandum. We shall not make it a necessary condition for a sound explanation, however, that the explanans must contain at least one statement which is not a law; for, to mention just one reason, we would surely want to consider as an explanation the derivation of the general regularities governing the motion of double stars from the laws of celestial mechanics, even though all the statements in the explanans are general laws.

(R3) *The explanans must have empirical content; i.e., it must be capable, at least in principle, of test by experiment or observation. This condition is implicit in (R1); for since the explanandum is assumed to describe some empirical phenomenon, it follows from (R1) that the explanans entail at least one consequence of empirical character, and this fact confers upon it testability and empirical content. But the point deserves special mention because . . . certain arguments which have been offered as explanations in the natural and in the social sciences violate this requirement.*

II. Empirical Condition of Adequacy

(R4) *The sentences constituting the explanans must be true. That in a sound explanation, the statements constituting the explanans have to satisfy some condition of factual correctness is obvious. But it might seem more appropriate to stipulate that the explanans has to be highly confirmed by all the relevant evidence available rather than that it should be true. This stipulation, however, leads to awkward consequences. Suppose that a certain phenomenon was explained at an earlier stage of science, by means of an explanans which was well supported by the evidence then at hand, but which had been highly disconfirmed by more recent empirical findings. In such a case, we would have to say that originally the explanatory account was a correct explanation, but that it ceased to be one later, when unfavorable evidence was discovered. This does not appear to accord with sound common usage, which directs us to say that on the basis of the limited initial evidence, the truth of the explanans, and thus the soundness of the explanation, had been quite probable, but that the ampler evidence now available made it highly probable that the explanans was not true, and hence that the account in question was not—and had never been—a correct explanation. (A similar point will be made and illustrated, with respect to the requirements of truth for laws. . . .)*

Some of the characteristics of an explanation which have been indicated so far may be summarized in the following schema:

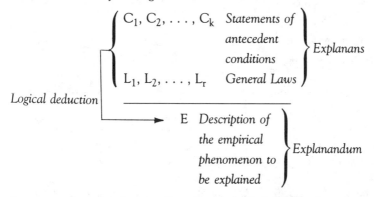

Let us note here that the same formal analysis, including the four necessary conditions, applies to scientific prediction as well as to explanation. The difference between the two is of a pragmatic character. If E is given, i.e., if we know that the phenomenon described by E has occurred, and a suitable set of statements $C_1, C_2, \ldots, C_k, L_1,$ $L_2, \ldots, L_r$ is provided afterwards, we speak of an explanation of the phenomenon in question. If the latter statements are given and E is derived prior to the occurrence of the phenomenon it describes, we speak of a prediction. It may be said, therefore, that an explanation is not fully adequate unless its explanans, if taken account of in time, could have served as a basis for predicting the phenomenon under consideration. Consequently, whatever will be said in this article concerning the logical characteristics of explanation or prediction will be applicable to either, even if only one of them should be mentioned.

It is this potential predictive force which gives scientific explanation its importance: only to the extent that we are able to explain empirical facts can we attain the major objective of scientific research, namely, not merely to record the phenomena of our experience, but to learn from them, by basing upon them theoretical generalizations which enable us to anticipate new occurrences and to control, at least to some extent, the changes in our environment.

Many explanations which are customarily offered, especially in prescientific discourse, lack this predictive character, however. Thus it may be explained that a car turned over on the road "because" one of its tires blew out while the car was traveling at high speed. Clearly, on the basis of just this information, the accident could not have been predicted, for the explanans provides no explicit general laws by means of which the prediction might be effected, nor does it state adequately the antecedent conditions which would be needed for the prediction. The same point may be illustrated by reference to W. S. Jevons' view that every explanation consists in pointing out a resemblance between facts, and that in some cases this process may require no reference to laws at all and "may involve nothing more than a single identity, as when we explain the appearance of shooting stars by showing that they are identical with portions of a comet." But clearly, this identity does not provide an explanation of the phenomenon of shooting stars unless we presuppose the laws governing the development of heat and light as the effect of friction. The observation of similarities has explanatory value only if it involves at least tacit reference to general laws.

In some cases, incomplete explanatory arguments of the kind here illustrated suppress parts of the explanans simply as "obvious"; in other cases, they seem to involve the assumption that while the missing parts are not obvious, the incomplete explanans could at least, with appropriate effort, be so supplemented as to make a strict derivation of the explanandum possible. This assumption may be justifiable in some cases, as when we say that a lump of sugar disappeared "because" it was put into hot tea, but it is surely not satisfied in many other cases. Thus, when certain peculiarities in the work of an artist are explained as outgrowths of a specific type of neurosis, this observation may contain significant clues, but in general it does not afford a sufficient basis for a potential prediction of those peculiarities. In cases of this kind, an incomplete explanation may at best be considered as indicating some positive correlation between the antecedent conditions adduced and the type of phenomenon to be explained, and as pointing out a direction in which further research might be carried on in order to complete the explanatory account.

The type of explanation which has been considered here so far is often referred to as causal explanation. If E describes a particular event, then the antecedent circum-

stances described in the sentences C_1, C_2, . . . , C_k *may be said jointly to "cause" that event, in the sense that there are certain empirical regularities, expressed by the laws* L_1, L_2, . . . , L_r, *which imply that whenever conditions of the kind indicated by* C_1, C_2, . . . , C_k *occur, an event of the kind described in E will take place. Statements such as* L_1, L_2, . . . , L_r, *which assert general and unexceptional connections between specified characteristics of events, are customarily called causal, or deterministic, laws. They are to be distinguished from the so-called statistical laws which assert that in the long run, an explicitly stated percentage of all cases satisfying a given set of conditions are accompanied by an event of a certain specified kind. Certain cases of scientific explanation involve "subsumption" of the explanandum under a set of laws of which at least some are statistical in character. Analysis of the peculiar logical structure of that type of subsumption involves difficult special problems. The present essay will be restricted to an examination of the causal type of explanation, which has retained its significance in large segments of contemporary science, and even in some areas where a more adequate account calls for reference to statistical laws.*

Has use of the covering law model simply forced us to return to the thesis of Chapter 3? Is it another way to embrace once again the notion of cause as the core of the meaning of explanation? In one way, yes. Hempel and Oppenheim refer to the deductive procedure as providing a causal explanation. Yet note carefully what they mean by "cause" in their analysis. We are entitled to say that an event is caused by an antecedent condition when laws assert that the event will occur whenever the condition is present. Compared to the range of formulations of the causal principle that we considered in Chapter 3, this definition is relatively weak. An antecedent condition does not need to have some "power" to produce the event, or at least we do not need to know of any such power, in order to identify it as a cause. To qualify as a cause, the antecedent condition need only regularly *precede* the event, and the law used is the assertion of the regularity of this relationship.

If we want to insist on a strong idea of causal connection—one that identifies the power in the cause that produces an effect—we must make claims to know more, to "see into things more deeply," than the covering law model asserts. This is one reason why positivists have often favored the covering law model. They may affirm that it is the structure of explanation, while insisting that the covering laws themselves are merely generalizations from experience.

The appeal of the covering law model is broad because it can be rendered consistent either with a positivistic position or with a formal cause view of explanation. The positivist can say that while the model does extend the meaning of explanation beyond mere description, still the covering laws themselves are only the results of descriptions. However, the advocate of a formal cause view may interpret the covering laws as more powerful than mere summary descriptions, and therefore as providing the backbone for a logically tight system of explanations.

Hempel and Oppenheim defend two theses that, when combined, lead to an interesting result. The first thesis is that the conclusion must be a logical consequence of a set of true statements. This means that there must be a necessary connection between the antecedent conditions and the conclusion, as claimed by the law cited in the explanation. But do we ever *know* of the existence of such a necessary connection? According to the second thesis, we do not. This second thesis is that our knowledge is only of observational regularities. One would defend this thesis by agreeing with the empiricists that all of our knowledge comes from experience, and with Hume when he said that experience never provides evidence of a necessary connection.

When the first and second theses are combined, the result is that no candidate explanation can be *known* really to be an explanation. This may sound odd, but it is just another way of saying that science cannot ever be sure it has reached the truth. It still may make progress toward explanations—gathering more and more phenomena under the umbrella of possible laws with wider and wider applications.

In some ways, the covering law model of scientific explanation finds its sources in Euclid and Descartes. Euclid's geometry is a model of deductive rigor, beginning from a few axioms (paralleling the covering laws) and proceeding, through the use of only the laws of logic, to expand these axioms into the many theorems of plane geometry. In a sense, all the theorems are "contained" in the axioms from the beginning. Rather than adding new information, the deductive system only makes explicit, by means of definitions and argument chains, consequences that will follow from the content of the axioms themselves. Logicians call such arguments analytic, meaning that the succession of theorems come not from adding new axioms or conducting experiments but from analyzing the content of the original axioms. As a model for science, this ideal of a deductive system does not demand that one abandon experiment. It *would* demand that experimental results do not become part of a science until they are connected to an axiom system of explanation.

Descartes, as we noted in Chapter 3, emphasized the formal (mathematical/logical) character of explanation and dreamed of a physics—indeed an entire natural science—that would be as deductively rigorous as Euclid's geometry. The covering law model can be interpreted as a modern restatement of that dream. Indeed, the model has gained such wide recognition in recent years that many commentators refer to it simply as the "standard account" of scientific explanation. Even for those who find it inadequate, it has provided the starting point of debate for four decades.

The positivists believed that *their* analysis of explanation was of immediate and great practical significance for scientists. It would, above all, prevent scientists from looking for answers to nonexistent questions and tolerating answers that demand that one subscribe to some particular metaphysical system. Causalists recommend *their* position, too, as aiding the practicing scientist, by showing where to look for explanations (mathematics, mechanical entities, or whatever is suggested by emphasis on one of the four Aristotelian causes). Is there corresponding practical significance in the covering law model? Certainly its proponents would say yes. Understanding it provides one with a logical "picture" of what an entire science looks like and how far it has progressed. The deductive chains provide a view of various levels of laws, all the way from universal "fundamental" laws, to experimental laws, to the phenomena that the science claims ultimately to explain. Attention to its demands will lead one to seek deductive connections between laws and stimulate one to seek connections between the laws of one science and the phenomena of another. Surely, it would be argued, it is important to show science as a hypothetical-deductive system, since the system exhibits the clear logical relationship between various levels of laws and between laws and events. For example, experimental laws in a thoroughly studied field of science may be shown to be the deductive consequences of more general laws. Fitting an experimental law into a larger framework by showing its logical place is called justification of a law. The practicing scientist is always looking for this logical framework for scientific laws.

Criticism of the Covering Law Model

Justification, and Discovering in Science.   The covering law model gives a clear explanation of how experimental laws are *justified*. We justify them by deducing them from covering laws. We noted, at the conclusion of the preceding section, that this strength is seen as of great practical importance by the model's defenders. Yet critics of the model point out that we need to know more about explanation than merely what justifies experimental laws. The neat, orderly arrangement of experimental and theoretical laws pictured by the covering law theory hardly reflects the practice of actual research. In a first laboratory science course, the experiments may be laid out clearly, with the relevance of the anticipated results already established (as confirming some set of experimental laws, within an assumption context). But when one is doing original research, things are much less tidy, as is clear when we realize that breakthroughs in research often raise questions about which assumptions were valid and how the results are to be interpreted.

To account for this difference between *learning* the body of a science and *doing* it, many contemporary philosophers of science have distinguished between the logic of justification and a possible logic (or form) of *discovery*. Whether there is one best logical procedure for discovering laws and acceptable theories is not clear, though many thinkers have tried to formulate one.[4] What is clear is that scientists, especially the ones who have made the most exciting discoveries, cannot easily be shown to have followed a deductive model. The covering law theory does not address such questions as "Where do hypotheses come from?" and "Are imaginative models of the relationships expressed by a theory necessary?" Presenting a body of scientific knowledge as a set of deductions in which all that is held to be true merely follows from axioms seems too simple, or somehow incomplete. Science in practice seems more empirical, more open than plane geometry, which *is* a deductive, analytic, tautologous enterprise. Mario Bunge summarizes this one-sidedness of the covering law model when it is claimed to be the *whole story* of explanation.

Unfortunately or not, logic does not tell the whole story of knowledge. Explanation, like deduction in general, does add to knowledge, because actually the object to be explained was not previously *contained* in its class (or in its law statement) from the start; it was put by us in it a posteriori. The explanation operation is not a mere drawing of an element out of a given collection; from an epistemological point of view, explanation does not consist in the mere identification of an element of a class that is overtly displaying its characteristics before us: explanation consists, rather, in the *inclusion* of the given object (fact or idea) in its class. And this is a *constructive*, synthetic operation, requiring the previous schematization of the given object, comparison of it with other objects, and so on. Now, at the level of formal logic, change does not enter; hence processes, like the epistemological one involved in explanation, have no place in deductive logic, which dispenses with the time concept and treats the actual processes of thought as laid down in an eternal present—just in order to avoid contradictions between successives. In other words, what from an epistemological viewpoint is a real transition from ignorance to knowledge appears in formal logic as a purely analytic relation. Deduction, and in particular explanation, always entails a novelty in knowledge—and this is why we care to perform it. To put aside the nonlogical aspects of explanation, by concentrating exclusively on its logical structure—as is usual among contemporary empiricists—is, I believe, a token of one-sidedness.[5]

We should note that Bunge's observations do not constitute adequate grounds for rejecting the covering law model itself. He has argued only that it must not be taken as an answer to all of the questions that scientists should ask about the nature and adequacy of explanations.

A covering law theorist might argue with the above criticism and claim that his or her model says everything that need be said about the adequacy of explanations. However, another option is available. The covering law theorist might, instead, agree with Bunge that the model is limited to providing the logical structure of justification. Even so, he or she could argue, this feature of explanation is important enough that the covering law model is an essential part of the answer to our central question, What makes one scientific explanation more adequate than another?

Scriven's Critique of the Covering Law Model.   The preceding criticisms could be made even by those who believe that the covering law model, properly limited in scope, is correct. But other critics charge that the problem is deeper—that the model is fundamentally mistaken.

In his article "Explanation, Predictions, and Laws," Michael Scriven presents arguments that may be summarized as follows: the deductive model may sound reasonable, but it's not the way things really are explained in science; the model presents us with an ideal that is not only unattainable but also confining, misleading, and therefore undesirable.

In the following excerpts,[6] Scriven presents some of the main areas of his disagreement with Hempel and Oppenheim.

EXPLANATIONS AS "MORE THAN" DESCRIPTIONS

[1] Another common remark in the literature is that explanations are more than descriptions. This is put by Hempel and Oppenheim in the following words: ". . . especially, scientific research in its various branches strives to go beyond a mere description of its subject matter by providing an explanation of the phenomena it investigates." But if one goes on to examine their own examples of explanations one finds what seem to be simply complex descriptions. Thus they offer an explanation of the fact that when "a mercury thermometer is rapidly immersed in hot water, there occurs a temporary drop of the mercury column, which is then followed by a swift rise." And the explanation consists of the following account: "The increase in temperature affects at first only the glass tube of the thermometer; it expands and thus provides a larger space for the mercury inside, whose surface therefore drops. As soon as by heat conduction the rise in temperature reaches the mercury, however, the latter expands, and as its coefficient of expansion is considerably larger than that of glass, a rise of the mercury level results."

[2] This is surely intended to be a narrative description of exactly what happens. The one feature which might suggest a difference from a "mere description" is the occurrence of such words as "thus," "however," and "results." These are reminiscent of an argument or demonstration, and I think partially explain the analysis proposed by Hempel and Oppenheim, and others. But they are not part of an argument or demonstration here, simply of an explanation; and they or their equivalents occur in some of the simplest descriptions. "The curtains knocked over the vase" is a description which includes a causal claim and it could equally well be put, style aside, as "The curtains brushed against the vase, thus knocking it over" (or ". . . resulting in it being knocked over"). The fact that it is an explanatory account is therefore not in any way a ground for saying it is not a descriptive account (cf. "historical narrative"). Indeed, if it was not descriptive of what happens, it could hardly be explanatory. The question we have to answer is how and when certain descriptions count as explanations. Explaining how fusion processes enable the sun to maintain its heat output consists exactly in describing these processes and their products. Explaining therefore sometimes consists simply in giving the right description. What counts as the right description? Tentatively we can consider the vague hypothesis that the right description is the one which fills in a particular gap in the understanding of the person or people to whom the explanation is directed. (pp. 174–175)

EXPLANATIONS AS "ESSENTIALLY SIMILAR" TO PREDICTIONS

[3] The next suggestion to be considered is a much more penetrating one, and although it cannot be regarded as satisfactory, the reasons for dissatisfaction are more involved. Quoting from Hempel and Oppenheim once more: "the same formal analysis . . . applies to scientific prediction as well as to explanation. The difference between the two is of a pragmatic character. . . . It may be said, therefore, that an explanation is not fully adequate unless . . . if taken account of in time, [it] could have served as a basis for predicting the phenomenon under consideration."

[4] *The full treatment of this view will require some points that will only be made later in the paper; but we can begin with several rather weighty objections. First, there certainly seem to be occasions when we can predict some phenomenon with the greatest success, but cannot provide any explanation of it. For example, we may discover that whenever cows lie down in the open fields by day, it always rains within a few hours. We are in an excellent position for prediction, but we could scarcely offer the earlier event as an explanation of the latter. It appears that explanation requires something "more than" prediction; and my suggestion would be that, whereas an understanding of a phenomenon often enables us to forecast it, the ability to forecast it does not constitute an understanding of a phenomenon. (pp. 176–177)*

[5] *In the primary use of explanation, then, we know something when we are called on for an explanation that we do not know when called on for a prediction, viz., that the event referred to has occurred: This is sometimes a priceless item of information since it may demonstrate the existence or absence of a hitherto unknown strength of a certain power. Thus, to take a simpler example than the bridge case, a man in charge of an open-hearth furnace may be suspiciously watching a roil on the surface of the liquid steel, wondering if it is a sign of a "boil" (an occasionally serious destructive reaction) on the furnace lining down below or just due to some normal oxidizing of the additives in the mixture. Suddenly, a catastrophe: the whole charge drops through the furnace lining into the basement. It is now absolutely clear that there was a boil which has eaten through the lining: apart from sabotage (easily disproved by examination) there's no other possibility. But no prediction is possible to the event, using the data then available. This renders almost empty Hempel and Oppenheim's (and even Scheffler's) conclusion that explanations provide a basis for predictions. For "Had we known what was going to happen, we could have predicted it" is a vacuous claim. One might mutter something about "If the furnace was in exactly the same state again we could predict it would dump," but I have already pointed out that this is a virtually empty remark since we usually can't identify "exactly the same state"; it is simply a dubious determinist slogan, not even a genuine conditional prediction. Since it is technically entirely impossible to rebuild the furnace to the point where it is identical down to the temperature distribution in the mixture (a crucial factor) and the shape of the irregularities in the floor (also crucial), even if we knew these specifications, it will be pure chance if the conditions ever recur and when they do they won't be identifiable. Thus our grounds for thinking the determinist's slogan to be true—if we do—are entirely indirect, and the explanation certainly does not rest on subsumption under the slogan since we cannot even tell when the latter applies, whereas we can be sure the explanation is correct. (pp. 188–189)*

EXPLANATIONS AS SETS OF TRUE STATEMENTS

[6] *It is not possible to claim that explanations can only be offered for events that actually occur or have occurred. They can be given for events in the future (Scheffler), for events in fiction, for events known not to occur, and for events wrongly believed to occur—and also for some laws, states, and relationships which are timeless. Assuming Hempel and Oppenheim's analysis to be in other respects correct, it follows that in*

such cases some of the propositions comprising the explanation itself cannot be true, contrary to one of their explicit conditions. The reason they give for this condition is a very plausible one, however, and it is of interest to see if a more general account can be given which will contain allowance for their point. They say, "it might seem more appropriate to stipulate that the [explanation] has to be highly confirmed by all the relevant evidence rather than that it should be true. This stipulation, however, leads to awkward consequences. Suppose that a certain phenomenon was explained at an earlier stage of science by means of an [explanation] which was well supported by the evidence then at hand, but which had been highly disconfirmed by more recent empirical findings. In such a case, we would have to say that originally the explanatory account was a correct explanation, but that it ceased to be one later, when unfavorable evidence was discovered. This does not appear to accord with sound common usage, which directs us to say that . . . the account in question was not—and had never been—a correct explanation."

[7] . . . The proper way of avoiding Hempel and Oppenheim's powerful argument is, I think, very simple; the secondary uses of "explanation" are legitimate but there are no such secondary uses of "correct explanation," the term which they substitute halfway through the argument. Remove the qualifying adjective "correct" and you will see that the argument is no longer persuasive. For consistency, this term must be and can be added to the occurrences of "explanation" in the premises. Overwhelming counterevidence does not necessarily lead us to abandon or even to put quotes around "explanation," but, as the argument rightly says, it does lead us to abandon the application of the term "correct explanation" (or "the explanation" which is often used equivalently). Hence we should regard Hempel and Oppenheim's analysis as an analysis of "correct explanation" rather than of "explanation," or "an explanation," and this is surely what they were most interested in. "Explanations," or "an explanation," or "his explanation," or "a possible explanation," do not always have to be true (or of the appropriate type, or adequate); they only need high confirmation, at some stage.

[8] Doesn't the notion of confirmation come into the analysis of "correct explanation" at all? It is not part of the analysis, which only involves truth; but it is our only means of access to the truth. We have not got the correct explanation unless it contains only true assertions, but if we want to know which explanation is most likely to meet that condition, we must select the one with the highest degree of confirmation. Good evidence does not guarantee true conclusions but it is the best indicator, so we need no excuse for appealing to degree of confirmation. Moreover, we have no need to adopt the skeptic's position that all possibility of knowing when we have a correct explanation is by now beyond reasonable doubt, and to restrict "knowing" to cases of absolute logical necessity is to mistake the empty glitter of definitional truth for the fallible flame of knowledge. (pp. 190–192)

THE DISTINCTION BETWEEN EXPLANATIONS AND THE GROUNDS FOR EXPLANATIONS

[9] It is certainly not the case that our grounds for thinking a plain descriptive statement to be true are part of the statement itself; no one thinks that a more complete

analysis of "Gandhi died at an assassin's hand in 1953" would include "I read about Gandhi's death in a somewhat unreliable newspaper" or "I was there at the time and saw it happen, the only time I've been there, and it was my last sabbatical leave so I couldn't be mistaken about the date," etc. Why, then, should one suppose that our grounds for (believing ourselves justified in putting forward) a particular explanation of a bridge collapsing, e.g., the results of our tests on samples of the metal, our knowledge about the behavior of metals, eyewitness accounts, are part of the explanation? They might indeed be produced as part of a justification of (the claim that what has been produced is) the explanation. But surely an explanation does not have to contain the evidence on which it is based. Yet, the deductive model of explanation requires that an explanation include what are often nothing but the grounds for the explanation. (pp. 196–197)

[10] When we say that a perfectly good explanation of one event, e.g., a bridge collapsing, may be no more than an assertion about another event, e.g., a bomb exploding, might it not plausibly be said that this can only be an explanation if some laws are assumed to be true, which connect the two events? After all, the one is an explanation of the other, not because it came before it, but because it caused it. In which case, a full statement of the explanation would make explicit these essential, presupposed laws.

[11] The major weakness in this argument is the last sentence; we can put the difficulty again by saying that, if completeness requires not merely the existence but the quoting of all necessary grounds, there are no complete explanations at all. For just as the statement about the bomb couldn't be an explanation of the bridge collapsing unless there was some connection between the two events, it couldn't be an explanation unless it was true. So, if we must include a statement of the relevant laws to justify our belief in the connection, i.e., in the soundness of the explanation, then we must include a statement of the relevant data to justify our belief in the claim that a bomb burst, on which the soundness of the explanation also depends. (pp. 197–198)

[12] Perhaps the most important reason that Hempel and Oppenheim have for insisting on the inclusion of laws in the explanation is what I take to be their belief (at the time of writing the paper in question) that only if one had such laws in mind could one have any rational grounds for putting forward one's explanation. This is simply false as can be seen immediately by considering an example of a simple physical explanation of which we can be quite certain. If you reach for a cigarette and in doing so knock over an ink bottle which then spills onto the floor, you are in an excellent position to explain to your wife how that stain appeared on the carpet, i.e., why the carpet is stained (if you cannot clean it off fast enough). You knocked the ink bottle over. This is the explanation of the state of affairs in question, and there is no nonsense about it being in doubt because you cannot quote the laws that are involved, Newton's and all the others; in fact, it appears one cannot here quote any unambiguous true general statements, such as would meet the requirements of the deductive model.

[13] The fact you cannot quote them does not show they are not somehow involved, but the catch lies in the term "involved." Some kind of connection must hold, and if we say this means that laws are involved, then of course the point is won. The suggestion

is debatable, but even if true, it does not follow that we will be able to state a law that guarantees the connection. The explanation requires that there be a connection, but not any particular one—just one of a wide range of alternatives. Certainly it would not be the explanation if the world was governed by antigravity. But then it would not be the explanation if you had not knocked over the ink bottle—and you have just as good reasons for believing that you did knock it over as you have for believing that knocking it over led to (caused) the stain. Having reasons for causal claims thus does not always mean being able to quote laws. (pp. 198–199)

[14] We may generalize our observations in the following terms. An explanation is sometimes said to be incorrect or incomplete or improper. I suggest we pin down these somewhat general terms along with their slightly more specific siblings as follows. If an explanation explicitly contains false propositions, we can call it incorrect or inaccurate. If it fails to explain what it is supposed to explain because it cannot be "brought to bear" on it, e.g., because no causal connection exists between the phenomenon as so far specified and its alleged effect, we can call it incomplete or inadequate. If it is satisfactory in the previous respects but is clearly not the explanation required in the given context, either because of its difficulty or its field of reference, we can call it irrelevant, improper, or inappropriate.

[15] Corresponding to these possible failings there are types of defense which may be relevant. Against the charge of inaccuracy, we produce what I shall call truth-justifying grounds. Against the charge of inadequacy, we produce role-justifying grounds, and against the complaint of inappropriateness, we invoke type-justifying grounds. To put forward an explanation is to commit oneself on truth, role, and type, though it is certainly not to have explicitly considered grounds of these kinds in advance, any more than to speak English in England implies language-type consideration for a lifelong but polylingual resident Englishman.

[16] The mere production of, for example, truth-justifying grounds does not guarantee their acceptance, of course. They may be questioned, and they may be defended further by appeal to further evidence; we defend our claim that a bomb damaged a bridge by producing witnesses or even photographs taken at the time; and we may defend the accuracy of the latter by producing the unretouched negatives and so on. The second line of defense involves second-level grounds, and they may be of the same three kinds. That they can be of these kinds is partly fortuitous (since they are not explanations of anything) and due to the fact that the relation of being-evidence-for is in certain ways logically similar to being-an-explanation-of. In each case, truth, role, and type may be in doubt; in fact this coincidence of logical character is extremely important. We notice, however, that there is no similarity of any importance between these two and being-a-prediction-of, where truth is not relevant in the same way, role is wholly determined by time of utterance and syntax, and only the type can be—in some sense—challenged. (pp. 200–201)

[17] It is also clear that calling an explanation into question is not the same as— though it includes—rejecting it as not itself explained. Type justifying involves more than showing relevance of subject matter, i.e., topical and ontological relevance; it involves showing the appropriateness of the intellectual and logical level of the content;

a proposed explanation may be inappropriate because it involves the wrong kind of true statements from the right field, e.g., trivial generalizations of the kind of event to be explained, such that they fulfill the deductive model's requirements but succeed only in generalizing the puzzlement. One cannot explain why this bridge failed in this storm by appealing to a law that all bridges of this design in such sites fail in storms of this strength (there having been only two such cases, but there being independent evidence for the law, not quoted). This might have the desirable effect of making the mainte-nance boss feel responsible, but it surely does not explain why this bridge (or any of the other bridges of the same design) fails in such storms. It may be because of excessive transverse wind pressure, because of the waves affecting the foundations or lower members, because of resonance, etc.

[18] So mere deduction from true general statements is again seen to be less than a sufficient condition for explanation; but what interests us here is that our grounds for rejecting such an explanation are not suspicions about its truth or its adequacy, which are the usual grounds for doubting an explanation, but only its failure to explain. Certainly it fails to explain if incorrect or inadequate, but then one feels it fails in a genuine attempt, that the slip is then between the cup and the lip; whereas irrelevance of type is a slip between the hand and the cup—the question of it being a sound explanation never even arises. One may react to this situation by declaring with Hempel and Oppenheim that the only logical criteria for an explanation are correct-ness and adequacy, the matter of type being psychological; or, as I think preferable, by saying that the concept of explanation is logically dependent on the concept of under-standing, just as the concept of discovery is logically dependent on the concept of knowledge-at-a-particular-time. One cannot discover what one already knows, nor what one never knows; nor can one explain what everyone or no one understands. These are tautologies of logical analysis (I hope) and hardly grounds for saying that we are confusing logic with psychology.

[19] Having distinguished the types of difficulty an explanation may encounter, one can more easily see there is no reason for insisting that it is complete only if it is armed against them in advance, since (i) to display in advance one's armor against all possible objections is impossible and (ii) the value of such a requirement is adequately retained by requiring that scientific explanations be such that scientifically sound defenses of the several kinds indicated be available for them though not necessarily embodied in them. Since there is no special reason for thinking that true first-level role-justifying assumptions are any more necessary for the explanation than any others, it seems quite arbitrary to require that they should be included in a complete explanation; and it is quite independently an error to suppose they must take the form of laws. (pp. 203–204)

THE ALTERNATIVE ANALYSIS

[20] What is a scientific explanation? It is topically unified communication, the con-tent of which imparts understanding of some scientific phenomenon. And the better it is, the more efficiently and reliably it does this, i.e., with less redundancy and a higher overall probability. What is understanding? Understanding is, roughly, organized

knowledge, i.e., knowledge of the relations between various facts and/or laws. These relations are of many kinds—deductive, inductive, analogical, etc. (Understanding is deeper, more thorough, the greater the span of this relational knowledge.) It is for the most part a perfectly objective matter to test understanding, just as it is to test knowledge, and it is absurd to identify it with a subjective feeling, as have some critics of this kind of view. So long as we give examinations to our students, we think we can test understanding objectively. (On the other hand, it is to be hoped and expected that the subjective feeling of understanding is fairly well correlated with real understanding as a result of education.) (pp. 224–225)

Scriven's specific arguments and illustrations speak for themselves, but it may prove worthwhile to summarize the main points of his analysis.

First, it is incorrect to say, as Hempel and Oppenheim do, that explanations are always more than descriptions. Sometimes that's exactly what they are (paragraphs 1, 2).

Second, it is incorrect to identify explanations and predictions. Predictions are often made when no explanation is available, and we sometimes can explain but are unable to predict (pars. 3, 4, 5).

Third, deduction can't characterize explanation generally because (1) deductive "explanations" are often not very explanatory, and are in fact sometimes trivial (pars. 16, 17); and (2) deductive form is often not necessary (pars. 1, 2, 12).

Fourth, far from having to deduce phenomena from laws, we may even cite cases where no appeal to laws is necessary at all (pars. 12, 13).

Fifth, it is confusing to demand that all explanations be true. We often ask whether an explanation is true, just as we may ask whether it is appropriate or whether it is helpful. Therefore, we must have recognized it as an explanation before we sought the grounds for saying that it is true. Asking for an explanation and asking for the grounds for an explanation are different activities. To demand that an explanation include its own grounds (the evidence for it, etc.) places a stultifying restriction on the giving of explanations in the first place.

Scriven's Alternative. Scriven clearly believes that very few things are right in the covering law theory. One may suspect, in such cases of pervasive criticism of individual points, that the critic believes the theory has gone wrong early, and on very fundamental points. Scriven does not believe that the covering law model is merely the wrong choice as the universal form of explanation. Rather, he believes that *any* theory that decides precisely what explanation will look like, in all contexts, will be misguided. The roots of his view are found, in our era, in the work of Ludwig Wittgenstein.

Wittgenstein, in his early work with Bertrand Russell, was interested in developing an unambiguous artificial (consciously created) language in which the meanings of all the terms were either simple or clearly definable in terms of simple ones.[7] But he later became persuaded that any such attempt was doomed to failure insofar as it tried to capture the richness of meaning of our ordinary language. Some of our most important concepts, he became convinced, are usefully vague, having a whole family of resembling, but not systematically relatable, meanings. When you tell a playful child to "stand roughly there," in order to take his picture, you don't have a clearly definable circle in mind. If his persistent teasing, by being just a little "off" in locating himself, forces you to draw a circle on the floor, that circle doesn't fulfill just what you had in mind in the original instructions; you didn't have *any* clear boundaries in mind. He has (and this is crucial) *not* forced you merely to *clarify* what you meant; he has forced you to *change* what you meant. Wittgenstein's point is that we do a disservice to our language, our knowledge, and our ability to investigate the world by trying to replace the richly nuanced vagueness of our ordinary language with the enforced neatness of an artificially clear one. Wittgenstein spent considerable effort on such basic terms as *seeing, reading,* and *understanding* to illustrate his point.[8]

Scriven's own proposal (pars. 2, 19) is that explanation is what fills gaps in the understanding. He is suggesting not merely an alternative definition but a different way of looking at explanation generally. The meaning of *explanation* depends heavily on the context, on the state of knowledge of the questioner, and on the many unstated (and perhaps unstatable) assumptions that affect the questioner—explainer situation.

Does Scriven propose the type of relativism which holds that scientific explanation is whatever anyone means it to be? Although Hempel has criticized the Scriven position on this score, Scriven doesn't believe such a theory commits us to a destructive subjectivity (individual relativism, cf. Chapter 4). His response is to ask whether teachers think that their tests, which supposedly seek to examine student understanding, are merely subjective. The point is not that we don't know what *understanding,* and consequently *explanation,* mean. It is rather that we may reflect an awareness of what *understanding* and *explanation* mean by the way we ask and answer questions, without being able systematically to define these terms (as deductions from covering laws or in any other way).

Defense of the Covering Law Model

As you might expect, proponents of the covering law model have not withdrawn from the field in the face of the above criticisms. One question at stake in this particular controversy is whether we should seek *change* of commonly accepted standards of explanation when they do not fit the model (such as deduction from laws and antecedent conditions) we believe most justified. Perhaps our task ought to be merely to understand, not reform, the kinds of explanations scientists give and receive. The covering law theorists advocate that those who accept nondeductive accounts of phenomena as adequate explanations should change their standards. They will say that science is successful only when it is clear and logically rigorous. The advantage of science over everyday accounts of nature is science's systematic character, in which logical and empirical criteria define the appropriate context for explaining. Covering law proponents are little concerned over the charge that their theory implies that there are no real explanations. Even an unattainable ideal of explanation, they counter, may guide scientific inquiry and keep us from endorsing confusion, vagueness, and contradiction.

The causalist, positivist, covering law, and ordinary language accounts we have considered all have strengths. Proponents for each are able to produce common examples of explanation that seem to render their particular view plausible. How does one go about choosing among them or developing an acceptable blend of them all? Perhaps it will help to dwell for a time on what they have in common. Are all of these accounts trying to answer the same question? In one sense, they are, and exploring this question will be the task of Chapter 6.

Supplementary Readings

Hospers, John. "What Is Explanation." In *Introductory Readings in the Philosophy of Science*, edited by E. D. Klemke, Robert Hollinger, and A. David Kline. Buffalo, N.Y.: Prometheus Books, 1980, pp. 87–103.

Here we have a good readable version of the covering law model with some consideration, though not a particularly friendly consideration, of objections. The introduction preceding the Hospers article is also worth reading.

Suppe, Frederick. *The Structure of Scientific Theories.* 2d ed. Urbana: University of Illinois Press, 1977.

What had seemed incontestable in the 1950s, Hempel's views on explanation, had become known as the "received view" by the 1970s. That meant it had moved to the same category as parental admonitions. Who says philosophy doesn't change? Suppe provides a lengthy introduction to a set of papers read at a 1969 symposium on the title of this book. It is not easy reading, but it is an important work.

Pears, David. *Ludwig Wittgenstein.* Cambridge, Mass.: Harvard University Press, 1986.

This is a thoughtful introduction to the ideas of Wittgenstein, respected enough to be reprinted after fifteen years with a new introduction. Understanding Wittgenstein is not easy but is worth trying, for no philosopher denies the importance of his ideas for twentieth-century philosophy.

Notes

1. Deduction and its counterpart, induction, are the two alternative ways in which we argue from premises to conclusions. Induction proceeds from the particular to the more general or from what is known from previous experience to the present or future. For example, a shoe falls to the floor, the earth and the moon are attracted to each other, and two large freely suspended masses show an attractive force; so we may judge that *all* masses are attracted to all other masses. Deduction proceeds the other way, from the general to the more particular. For example, the first law of thermodynamics holds that

heat travels from a hotter object to a colder object. Given the condition of a cold winter day and a warm house, we can conclude that the temperature of the house will go down and the surroundings will warm up a corresponding amount.

Inductions involve a leap of faith for not every particular has been observed. We may induce that all swans are white because every swan we have ever seen is white. That particular induction, long used in logic textbooks, turned out to be false when a variety of black swans was discovered. Deductions, on the other hand, are logically *necessary*. Their only potential weakness is that the premises might be false. A deduction does not require that the premises be true but only that the conclusion *follow* from the premises (if they are true, it is also).

2. It is also often called the hypothetical-deductive model when one wants to emphasize that the premises are hypotheses suggested by experience.

3. Carl G. Hempel and Paul Oppenheim, "Studies in the Logic of Explanation," *Philosophy of Science* 15 (1948): 135ff. Part 3 of the essay, not included here, presents a much more detailed attempt to work out the formal conditions for explanation. Those who are acquainted with symbolic logic would benefit from reading it.

4. Norwood Russell Hanson suggests the complexities, and possible futility, of such a task in his book *Patterns of Discovery* (Cambridge, England: Cambridge University Press, 1958).

5. Mario Bunge, *Causality* (New York: World Publishing Co., 1963), p. 289.

6. Michael Scriven, "Explanations, Predictions, and Laws," in *Scientific Explanation, Space, and Time*, vol. 3, *Minnesota Studies in the Philosophy of Science*, ed. Herbert Feigl and Grover Maxwell (Minneapolis: University of Minnesota Press, 1962). We have omitted the footnotes and references from Scriven's text and have numbered the paragraphs for convenience in referring back to them later. Page numbers are given after each excerpt.

7. Cf. his *Tractatus Logico-Philosophicus* (London: Routledge and Kegan Paul, 1961).

8. Cf. his *Philosophical Investigations*, trans. G. E. M. Anscombe (New York: Macmillan, 1953).

/6/ SCIENCE AND REALITY

Reality as the Object of Scientific Inquiry

Both the disagreements and the common ground of the alternative accounts of explanation run deeper than we have so far explored. The possibility of a natural world beyond human perception was an important source of controversy; and this fact alerts us that at stake are fundamental views concerning what reality is and how humans as knowers are related to it.

Where people disagree about important matters, however, will there not be some areas of consensus that provide the common context for argument? Where can such common ground be located for the positions we have examined? Hume, Mach, and Kant all denied the existence or the knowability of a reality beyond human perceptions. On what common ground could they agree with thinkers, like Descartes, who affirmed such a reality?

What these accounts do agree upon is that reality confronts humans regardless of their choice, their wishes, their values. While we may choose not to open our eyes, when we do open them we *discover*; we don't consciously *create* what we see. Our will does not make the world what it is, though our will may influence how we respond to it. In short, we *have* values, but we *learn* facts. Proper scientific inquiry will proceed by separating what we want to see from our research and explanation procedure.

Most of us will be sympathetic with this distinction between a world of fact on the one hand and human judgments of goodness and badness on the other. Indeed, we may wonder what alternative account would make sense. We will explore such an alternative in Chapter 7. For now, we should be clear on the sense in which the accounts considered so far share this common view.

Causalists like Descartes present us with clear cases of the affirmation of the fact/value distinction. Descartes urged that we not only keep our wishes out of our scientific inquiry but that we refrain from using judgments about God's purposes as well.[1]

Both the positivists and Kant base their analysis of nature on the realm of human experience. The positivists stress the ultimate character of sense impressions. Kant urged the recognition of the role of human classification in experience. For both, however, the fact that science depends on human experience does not mean that humans choose what they experience. The sort of subjectivity this dependence on choice would imply is denied by the positivist's reliance on a "given" in sense experience, and by the claimed universal human structure for experiencing in Kant.

The covering law theory clearly depends on our ability to recognize facts and to agree on them. The theory emphasizes the logical rigor and universal clarity of deductive argument. But suppose that when it came to identifying the "antecedent conditions" needed for the premise, everyone had his or her own interpretation of them and agreement could not be reached. In that case, the deduction would not yield uniform results, and the advantages of deduction would disappear.

Finally, Michael Scriven, the critic of the covering law model, also can support the fact-value distinction. He insists that his concept of "explanation as filling gaps in the understanding" is not dependent on variable individual judgment.

If all these theorists agree that science is getting at a reality that is independent of human choices, then judgments about what that reality is like become critical in assessing the theories. In this chapter, we will consider several of the traditional answers to this question and the compatibility of these answers with the several theories of explanation we have been examining.

Among the many questions that philosophers have asked about reality as a whole and our relation to it, two stand out as particularly important for the concept of explanation. The first is whether the universe is made up of individual entities alone, or whether "universals" exist as well. By "universals" we mean any quality or relationship that obtains over many individuals. For example, "water" is the name of a *kind* of molecule. Shall we say that only individual molecules exist, or does the *kind* of molecule exist as well? Solubility, liquidity, salinity are all general properties. Do they exist, in some sense, or is existence confined only to the individual entities that we label as soluble, or liquid, or saline? "To the right of," "larger than," and "dependent on" are names for relations between things. Do only the things related exist, or are the relations themselves real? If universals are not real but only human contrivances, then scientific laws (as one example of a universal), aren't statements corresponding to a reality either. Therefore such laws would not form an essential element in an explanation that attempted to reflect reality.

A second important question for us is whether the world is, at base, material through and through, or ideal, or some mixture of both. That is, is the universe composed of material things (chairs, or chemical elements, or subatomic particles), or is it also—or only—populated with entities such as perceptions, or ideas, or laws, or processes? If the universe is made up of both material and ideal entities, how are these two different sorts of things related to each other?

We will leave this second question, on the reality of material things and ideas, for later in this chapter. Now, we turn our attention to the first question, on the reality of universals.

The Reality of Universals

Wartofsky on Laws.   In the following selection from his book *Conceptual Foundations of Scientific Thought*, Marx Wartofsky explores the three classical answers that have been given to the question of the reality of universals. These three positions, realism, nominalism, and conceptualism, may be unfamiliar; however, several of the theories of explanation we have already discussed are reflected in those three answers. This is, of course, no accident. The main ways of identifying scientific explanations find their roots in these views on the nature of reality.

Wartofsky begins, appropriately, with an introductory discussion of laws, because one's view of the status and importance of scientific laws is directly related to these ontological positions.[2]

LAWS

[In earlier parts of my account,] the object of scientific inquiry appeared in large part to be the formulation of statements of universal scope which state some invariance between properties or events. Such a statement, of the form $(x)(Fx \rightarrow Gx)$, for example, formulates a law, asserting conditionally that if anything has the property F, then it also has the property G. From our earlier discussion it should be clear that this simple formulation conceals a number of problems. First, the import of the statement is that as a matter of fact, if anything has F it has G, unrestrictedly. Suppose that F is the property of being a day in the month of June, 1961, at some location L, and that G is the property of not registering a temperature lower than 45° F at L. The presumed law would then be asserted in the statement, "If anything is a day in the month of June, 1961, at L, then it will not register a temperature of less than 45° F at L. One could state this as a hypothesis to be confirmed by the evidence and then check the suitably defined "readings at L" for each of the thirty days in June of that year at L, and discover whether or not the law is true. The law would then be nothing but a summary statement not only of all the observed instances of x, but of all the possible observed instances of x. The condition of the law statement being a true statement is precisely that for every instance of x, what is asserted of it is true. We cannot know that the statement is a law unless we know whether it is true or not, or we would be in the peculiar position of stating "laws" that were sometimes true and sometimes false (or in the no less peculiar position of stating laws that were not known to be true or false). In the former case we might claim that the statement, "June 22, 1961, at L yields temperature readings of 78° F at L," is a law which is true from 2:30 P.M. to 3:15 P.M. and false at other times. In the latter case we might claim that it is a law that the

temperature on Ceres varies within a range of ± 247° F, but we do not know whether this law is true or false. But unless we knew whether or not this were true, it would be odd to assert it as a law. We save the situation by talking of it as a hypothesis which, if it is true, states a law; therefore, a "law-like statement." But the only sort of instance in which we could know that a law is true on these grounds is one like the "law" concerning the days of the month of June, in which all the possible observations can be made in a finite time, in a complete enumeration. But a simple enumeration of all known instances is not what we generally regard as a law, though it does state an invariance, and states it to hold universally over all the members of a class of facts. We require a law to state that something is unrestrictedly true for all possible instances where the number is presumably indefinitely larger than that of the observed instances, for we take a law to be a generalization which goes beyond the presently available evidence. Yet if it goes beyond the available evidence, we cannot know that it is true in all instances. We are thus in the unhappy position of saying that if a law is true, we cannot know it, and if we know that a universal statement is true for all of its instances then it is not a law. As "unhappy" as this may sound, it sets forth certain conditions for laws which are important in explicating our concept of law and serves to interpret the sense of a statement like (x)(Fx → Gx) when we take it to be a law.

In order to make this clearer, we may distinguish between "laws of nature" and the "laws of science." If we mean to preserve the sense of objectivity in the claim that something is a law of nature, we ought to consider that it is a law of nature whether anyone knows it or not. Thus, we assume that Galileo's law concerning the acceleration of falling bodies did not come into being as a law of nature when Galileo formulated it. Rather, we assume that this law holds true for any time at all, including the time before there were conscious beings on this planet. If laws of nature are capable of being discovered (that is, if it makes any sense at all to say a law is "discovered"), then the relations of invariance which the laws assert hold whether they are known or not. Further, we assume that the law will hold in future, or as yet unobserved instances, whether or not these instances in fact come to be observed or not. We may formulate this in an assertion that underwrites the empirical nature of the law; thus, "if any instance were to be observed, then it would conform to the law." This is precisely the assertion that the law holds in those instances which have, as a matter of fact, not been observed; and stronger yet, whether or not they are ever observed in fact. Because this raises, as a condition, the epistemologically impossible situation of there conceivably being observers before there were any conscious beings on the planet, we may say that such a contrary-to-fact assertion cannot be meant to assert the possibility of observation in such instances, and so even this condition of the empirical nature of the law seems too strong, if we want to maintain that Galileo's law held in fact prior to the appearance of conscious beings.

The claim we tend to make for a law of nature is that it hold independently of whether anyone knows it or not, and even independently of whether it is possible to know it (which is a good deal stronger). But this is the claim for its objectivity. For if we made any lesser claim than this, we would have to say that natural laws hold only for those instances which are known, or only for those instances which may, as a matter of factual possibility, come to be known. Thus, the factually unfulfillable epistemological condition, "if anyone were to observe a falling body prior to the appearance of conscious beings . . ." may be reduced to the ontological assertion, "If there were a falling body, then it would have accelerated in accordance with Galileo's law." Such an "if-then" conditional in the subjunctive mood ("were-would") is called a subjunctive conditional. There may in fact have been such falling bodies (and we assume there were); therefore such a conditional is not contrary to fact. But it is contrary to fact that there was an observer before the appearance of conscious beings, and in the epistemological form of the assertion, this would be called a "contrary-to-fact conditional" or a "counterfactual conditional." We may formulate this in an even stronger sense. We might say, even if there never was a falling body, and never will be, in fact; still, if there were, it would fall in accordance with Galileo's law, thus taking the law to hold in every possible world, regardless of whether it holds in any actual world.

We may also assert counterfactually in the case where we do know that something did not occur as a matter of fact, "If I had dropped the professor out of the window ten minutes ago, he would have accelerated in accordance with Galileo's law," where as a matter of fact, the antecedent is known to be false, because I did not do it (though this may also be said to be epistemologically dependent: "I know I did not do it, because I remember, or because there he is walking in the hall without a scratch, to all appearance"). Our sense of "law of nature" in this case is that the law is "real" or "out there" in the world or in nature, with or without my consent or even the possibility of my observing instances of it. Thus, the assertion of the law is an inference, either from certain evidence or from other laws from which it can be deduced. If it is an inference from certain evidence, then it is presumably an inductive inference; namely, one which if it is in accordance with certain norms, is taken as the ground for a rational or warranted belief. If it is deduced from other laws of greater generality, as a deductive consequence, then its force is no greater than that of the weakest of the premises from which it has been deduced. If it follows deductively from some premises taken to be necessarily true (e.g., synthetic a priori "truths," or metaphysical "truths," or logical "truths"), then I may claim that the law is also necessarily true; but observation would then play no role in confirming it, because it would be true independently of any observation and would thus be an a priori law, not an empirical law; i.e., it is not one which may be confirmed or falsified by test. My assertion of the law, if it is empirical at all, is at best a warranted assertion on the basis of evidence, if any of the premises from which it is deduced are in turn empirical generalizations taken to be universal statements warranted on the basis of evidence. Thus, I cannot know the law to be true as a law of nature, but I can know that my belief that it is true is a rational or warranted belief, on the basis of the evidence.

We may then say that the "laws of science" are hypotheses or postulates which are the objects of rational belief on the basis of evidence and that if in fact the laws of science are true, then they state laws of nature. This is a version of the "correspondence theory" of truth (or the "semantic theory of truth" which is formulated by Tarski and others with respect to statements in general, but not specifically with respect to laws or law-like statements). Where "L" stands for the law of science and L for the law of nature, this states that "L" is true if and only if L. Or, for example "s $= \frac{1}{2}gt^2$" is true if and only if, as a matter of fact, the distances which a freely falling body traverse stand to each other in the same ratio as the squares of the times of fall.

This may be called the realist view of the nature of laws of nature. You may be sure that not all philosophers are realists in this sense and that there are therefore alternative views. We will examine these next, under the headings nominalist and conceptualist views. But having introduced some initial ideas, let us summarize them briefly.

(1) A law states an invariant relation among all members of a given class (with respect to some parameters taken to be relevant). This may be in the form of a universal conditional—(x)(Fx → Gx)—or a biconditional—(x)(Fx ↔ Gx)—or in the stronger form of the subjunctive or counterfactual conditional: For any x, if Fx were the case then Gx would be the case. We take a law of nature to hold for an indefinitely large or infinite class of natural events or instances and, in general, to be unbound to any particular time, or to hold indifferently at any time. (As we shall see, this raises special problems concerning laws for historical epochs which are assumed to hold within a bounded time, or to be time variant, such as geological, biological, or sociological laws of a certain sort.)

(2) A law-like statement states a law of nature if it is in fact true for all the instances subsumed under the law, when these instances are of the sort set forth in paragraph (1), and thus the domain of the law is such as to support a genuine generalization, and not simply a summary description or list of all the instances. The conditions under which the law could be asserted to be true are expressed by the subjunctive-conditional form, "If, for any x, Fx were the case, then Gx would be the case." The subjunctive conditional leaves open the possibility that in fact something was, will be, or is the case, but implies that this is not known to be true. The epistemological condition of knowing that something was not or is not the case, as a matter of fact, combined with the assertion that if it were, the law would hold, expresses a belief that the law is true unrestrictedly and takes the form of a contrary-to-fact conditional or counterfactual conditional. The distinction between not knowing whether something was or was not the case and knowing that it was not the case distinguishes the subjunctive from the counterfactual conditional.

(3) A law of science is a statement of a law of nature which is not known to be true in all of its instances, but for which there are grounds for rational or warranted belief. (Thus, a law of nature is predictive: that is, it makes a claim about instances not yet known to be true. "Predictive," more restrictedly, relates to future instances; reference to past instances is "retrodictive," therefore.)

The realist view of laws is "realistic" in the technical sense which this term has in epistemology and ontology. In this view, a law is a universal and the relations of invariance which it asserts exist in nature independently of their being known, or of the conditions under which they come to be known. Knowledge of this universal is therefore a discovery of it, and belief that such a universal law exists, as a matter of fact, is warranted by the validity of the knowledge claim in terms of certain norms. In this context, at any stage of science the prevailing laws state warranted beliefs about the laws of nature, or about the truth of the propositions which state such laws. In another interpretation of this view, any scientific law is a perspectival or partial "truth," relative to the evidence and the framework within which the evidence is significant. Thus, the true propositions in which the laws of nature are asserted are objectively true, but our knowledge of them, at any time, is relative. The fallibility of laws of science lies, therefore, in this relativity, so that what was once a warranted belief turns out no longer to be so, when the evidence is increased, or when the framework within which the evidence is significant is abandoned and a new framework adopted. The reasons for such changes are in part the ones [discussed earlier]. In such a view, the constant "correction" of a statement of relative frequency, with each additional instance, would assume such a constant approximation toward a limit.

The realist view of laws thus assumes that there are objectively true propositions asserting such laws, to which scientific laws constantly approximate as alternative hypotheses are eliminated, or as evidence increases and as criticism and refinement of methods proceeds.

Thus, one sense of realism, which we may characterize as the epistemological sense, takes laws to be true or false objectively, independently of whether they are known to be true or false. Laws, therefore, are not simply ways in which we shape the world of experience to our knowledge but that to which our knowledge has in some way to conform, if it is veridical. In another sense of realism, related to the first in a complex way, laws of nature are taken to be "real universals" (in the Platonic tradition of the "forms") which exist (or subsist) objectively and are discovered by means of rational inquiry. Thus, the instances simply exemplify these universals, or are the means whereby these are revealed to us.

If the realist, in this sense, assumes that universal laws of nature exist in reality, the nominalist challenges the view that universals exist at all. The distinction in names comes from a medieval philosophical controversy on the nature of universals, the realist view asserting that Universalia sunt realia ("Universals are reals"), and the nominalist arguing that Universalia sunt nomina ("Universals are names"). The force of the nominalist's objection, as it has been interpreted, has a certain empiricist flavor: in our experience, we do not come to know universals, but only particulars (whether sense data, or particular instances or events). We collect such instances in accordance with common features which we recognize among them. Universals "exist" only in the common names which we use to mark these common features. They may be conveniently named in common, but the common features do not exist apart from the particular instances which we experience or which occur. Thus, the only status universals have is as names, and these names "exist" also only as particular marks, or inscriptions, or utterances on particular occasions. Thus universals do not exist. With respect to the account of laws, this asserts that laws "exist" in nature only in the instances in which certain features are exhibited. There is no universal connection among these features, except in the sense that they may all be gathered under a single expression, which serves conveniently to group them for purposes of reference. Thus, for example, "The people of the United States," does not denote a universal, but only a collection of individuals, who may be named in common. Only the name The people of the United States exists, and exists only in particular instances such as the one in this sentence, which is a particular inscription in ink on paper. Such a law as $s = \frac{1}{2}gt^2$ is also such a name, whose reference is not some universal "real," but only the collection of the particular instances in which a particular object at a particular time and at a particular place falls in such a way that its time of fall is in a certain ratio to the distance of fall.

There are major philosophical difficulties with both these views, realist and nominalist, which we cannot pursue here. But it may be recognized that this view lends itself to the interpretation that a law of science is only a convenient "mark," which serves a linguistic community as a way of denoting or sorting out a set of instances. The "convenient" or "economical" shorthand description which such law-like statements give is their only function. In this way, the nominalist tends to support an instrumentalist view of laws and theories, taking them simply as more or less adequate means of dealing with nature, rather than as true or false, as the realist sees them. But this function can be of great importance precisely because it facilitates and extends the range of human memory, inference, computation. It is this sense of "economy of thought" which Mach has in view, in the characterization he gives of mathematics, for example. But a conventional or convenient mark is neither true nor false. The sense in which a statement of relations is true or false, in the nominalist view of laws, is entirely comprehended in its "extension," i.e., in the instances to which it refers. Thus, many of the versions of "instantial confirmation" which mark empiricist approaches to hypotheses carry over this nominalist emphasis.

The conceptualist view of laws is an attempt to overcome difficulties in the nominalist view and in the realist view. If the nominalist says that laws are not real universals, but are "universal" only in the collection of instances marked by a common name or description, and if the realist claims that universals exist or are real, in nature or in the world, the conceptualist raises questions about each view. Does the realist mean that the universals exist apart from the instances which embody them? If so, then there is some realm of universals apart from the world of particular facts, and the relation of these universals to the world of facts presents insuperable difficulties. This creates a dualism of universals as "ideal forms," subsisting in some ideal realm (as in Platonic realism), and thus grants to the world of our experience or thought only a shadowy, secondhand reflected existence. If, on the other hand, the only real existents are particular facts, then the fact of their relation becomes an incomprehensible one, for a collection is no more than a collection unless there is some law-like relation which "really" holds among the particulars. (In Poincaré's phrase, "Science is built up of facts, as a house is built of stones; but an accumulation of facts is no more a science than a heap of stones is a house.") If the nominalist claims that he is arranging the facts under some conventional name or mark, then he disguises the contribution he makes in making the arrangement under the pretense of doing no more than naming. In effect, the statement of relations contributes something which the mere listing of accumulated particulars does not contain: the discovered order or relation among these particulars, or that in virtue of which they permit themselves to be grouped commonly. This is not explicit in the particular facts, as some real universal in which they are bound together, like raisins in a lump of dough. Rather, in discovering the relation, the mind makes the connections, explicating what is implicit in the particular facts. The universal is thus constructed in the mind, or is conceptualized as the order which is revealed by inquiry. In the Aristotelian version of this sort of conceptualism, the order or relation among the facts is potential, but is actualized by the process of conceptual discovery and construction. A law of nature is, so to speak, in nature, but not apart from the particular and concrete processes of nature. (The notion of process thus lends itself to this view more readily than the notion of discrete and isolated events, which already requires some glue to put together the pieces.) Thus, the universal or the law has no independent subsistence as an ideal form. By virtue of its implicitness in the real relations among events which constitute natural processes, the mind can bring this form or order into explicit consciousness, emulating or creating an ideal imitation of nature. Thus laws represent natural processes in the form in which these become known to a rational intelligence. Laws of science are therefore the form in which laws of nature come to be objects of reason, or of conceptual judgment. The laws of nature are realized (or actualized) in this conceptual activity, but are true or false because they do or do not adequately represent the lawful relations in nature. Thus, they are not conventions, although the form in which they are expressed may be conventional. Thus, the expressions $s = \frac{1}{2}gt^2$ and $AD/AE = (HM)^2/(HL)^2$, which are alternate ways of stating the same law or relation of physical invariance, are conventional, in that the choice of a language in which to express the relation is conventional. But both of them express the same proposition, and the proposition is not conventional, but true or false about relations among facts. The conceptualist might charge the realist with constructing abstract universals as if they were real things, thus hypostatizing a set of factual relations, in terms of their abstract expression. The nominalist avoids hypostatization at the expense of losing the objectivity or concreteness of the relations among facts. (The nominalist might answer that relations are also particulars in his scheme.) In laying claim to the best of both worlds, the conceptualist may then talk of laws as concrete universals, thus more than abstract universals and more than the mere aggregation of concrete particulars. In other versions, the conceptualist may claim that laws are universal forms imposed on experience as the condition under which experience comes to be known as more than undifferentiated flux. These conceptual forms are, in different versions, either a priori conditions of human knowledge (and thus conditions of the possibility of science), as a Kantian view might propose, or they are intuitions of an intellectual sort which immediately perceive the forms in terms of which experience comes to be known. These forms, then, constitute the rational structures themselves as the concretely presented features of nature, where the particular experiences are viewed as abstractions (abstracted from the whole, or the Gestalt, which is directly known in intuition and conceived of only abstractly as particulars, for the sake of analysis, or as explication of what is already known).

Within each of these philosophical positions the relation of universals to particulars is conceived differently; and in this context, the relation of laws to their instantiations is conceived differently. It is clear that this will have a bearing on the sense in which laws will be taken as explanations of particular events or occurrences.

In the realist view, a particular event is explained by a law, in the sense that the fact is an instance of the law, and can be shown to "follow" from the law: that is, the particular fact finds its place in the systematic relation which the law asserts.

For the nominalist, because there is no underlying "reality" beyond the facts themselves, the "law" is only a convenient shorthand description or summary of the particular facts, and thus laws cannot be said to explain at all. In this sense Bees give honey, does not explain why a particular bee gives honey, or why any particular bees whatever give honey; it is only another way of saying the same thing, and at most gives a description which is true of all the instances. But true here means no more than the *adequacy of the description, in its use as a mark understood by some linguistic convention as standing for the whole list of statements: Bee 1 gives honey, Bee 2 gives honey, Bee 3 gives honey . . . just as The People of the United States stands for the list containing the proper names of each individual in the United States (or each citizen, or each voting adult). The proper listing is thus a matter of a convention as to what The People of the United States will mean. In legal documents, for example, the description The party of the first part is explicitly defined, and thenceforward, the descriptive phrase stands for all and only those individuals for which it is explicitly defined.*

For the conceptualist a law explains in the sense that what is implicit in experience is brought into conscious and explicit form as an object of the understanding. Thus, the scientific law explains by virtue of realizing or exhibiting to conscious thought what was previously implicit in experience, or "in nature." Explaining and bringing to conscious understanding or to explicit conceptual formulation are thus one and the same, so that explanation becomes a matter of explication, of making explicit to the understanding. One variant of this view is that explanation consists of interpreting or translating what is unfamiliar into terms which are already familiar, as in the instances of anthropomorphic explanation we have discussed. This is also characteristic of the use of models and analogies, as aids to understanding, because the model or the analogy converts an unfamiliar set of relations into the form of some familiar configuration, as the Ping-Pong ball model of chain reaction, or the analogy of "lines of force" as strings under tension "makes clear" in more familiar terms what sorts of relations obtain in chain reactions and with respect to "lines" of force (where line is already an analogy, to geometric concepts).

This account of alternative positions is not to be taken as a literal distinction establishing cubbyholes into which alternative views of laws fit neatly, because any such position in the "pure" form is probably not to be found. Rather, these distinctions indicate emphases, and help to explain how these emphases give rise to or reflect alternative views of the nature of laws.

Wartofsky's account reflects the standard classification of positions concerning the reality of universals. Realism holds that universals (qualities, groups, and relationships) exist in their own right, apart from the particulars that have them, or participate in them. The law of gravitation, for example, might exist even if there were no bodies to be subject to it at a given time. Conceptualism is the view that universals exist, but never apart from their expression in particulars. The law of gravitation exists, in this view, just as fully as do the bodies related according to it. It would not exist in any sense, however, without the existence of the particular bodies. According to nominalism, only particulars exist. The law of gravitation, a universal, does not exist but is only a humanly devised way to summarize how those individual bodies we experience behave.

The three positions discussed find expression in the theories of explanation dealt with earlier. Realism provides the foundation for asserting the causal relations exist in nature. Nominalism finds its modern defense in the positivists. One position to which conceptualism can lead is Kant's defense of the causal organization of human experience.

A Review of the Positions: Strengths and Weaknesses. The importance of the problem of universals for the concept of explanation suggests that we explore further the strengths and weaknesses of the approaches Wartofsky has outlined.

The most fundamental attraction of *realism* is its insistence that a world of laws and other relationships exists independently of man's creations. We do not ourselves order the world; it is ordered already. Some realists point to the element of surprise in scientific inquiry, which comes when we discover an order that we did not at all expect. Others suggest that the faith that such an order exists is an essential element in the drive for scientific discovery. In this vein, James Watson spoke of the discovery of the structure of the DNA molecule. Speaking derisively of the empiricist demand for confirming tests that others might make, Watson said he and his associate Crick knew that the structure they had come upon was the right one because it was so beautiful. Experimental confirmation seemed hardly necessary to Watson because his model fulfilled so well his expectations for what relationships in nature should be like—elegant and simple.[3]

Associated with this advantage of realism is its ability to make sense of counterfactuals and subjunctive conditionals. These "if . . . then" statements are common and *seem* meaningful. Consider the following examples. "If you had studied your chemistry, you would have passed the test." "Were water to be found on Mars, it would be composed of hydrogen and oxygen in the ratio of two atoms to one." The first of these statements is a counterfactual, meaning that its premise, the "if" clause, is contrary to fact. The second is a subjunctive "if . . . then," or conditional, statement. In the first case, the particular event (studying), about which the universal claim (studying aids success) is made, does not exist. In the second case, nothing is said about whether there is, in fact, water on Mars, but, "If there is, then. . . ." Now if laws of nature and other universal relationships exist only when the individuals that they relate exist, then both of these propositions must be either false or, even more likely, meaningless. Yet they seem to have meaning, and at least the second surely is true.

Many empiricists, however, believe realism to have a fatal flaw. If all of our knowledge comes from experience, and if our experience is of particular sensations, then how can we ever assert knowledge of universal relations that are not experienced? And if some of our knowledge does *not* come from experience, where does it come from? From thought? But thought itself seems so often (perhaps always) to deal with abstractions from sense experience (note Mach's view on the nature of mathematics, cited in Chapter 4). What can account for any direct access that thought might have to reality, bypassing sense experience? Because of the attractiveness of realism's strengths, its proponents have spent a great deal of energy trying to answer these questions.

The two great strengths of *nominalism* are its simplicity and, in its empiricist form, its reliance on sense experience to which everyone agrees we have access.[4]

For philosophers through the ages, simplicity has been an alluring goal. Why? Because the truth of complex arguments seems dependent on simple arguments that are the components or elements of the complex; and simple arguments themselves are combinations of concepts that are simpler yet. Consider the argument:

(Major premise)   All events have causes.

(Minor premise)   The processes of the universe since its inception can be considered as one long event.

(Conclusion)   The universe has a cause.

There has been a great deal of disagreement over the truth of the conclusion cited above, and many philosophers have suspected that the disagreement may be the result of confusion caused by complexity and vagueness. Surely one would evaluate the soundness of the conclusion by considering the truth of each premise, and this would be a search by simplification. But the first premise, "All events have causes," is itself a difficult one, as we have learned. Our search for truth would perhaps lead us to analyze the terms *event*, *have*, and *cause*. Nor would our search likely end there, for we have seen that the word *cause* alone has a variety of senses, which we would need to scrutinize. Is there no end to this process? Can we ever hope to find the truth concerning complex arguments? It has been suggested that the only way to succeed is to reduce them to ultimately simple terms. These terms would refer to the simplest objects of knowledge or experience.

Nominalists insist that the simplest elements are certainly not universals such as "blue" and "horse," which refer to many instances of blue and to many individual horses. Simples, of course, aren't relations either—such as "iron atoms are bigger than hydrogen atoms"—because relations are obviously compounds of concepts. The heart of nominalism is the claim that the simples from which knowledge must start are particular, individual entities. Those nominalists who are also empiricists see these particulars as objects of sense experience. Hume called them impressions; Mach called them sensations; but the nominalist tone is clear in both. The great promise nominalists see in their program, then, is the achievement of as much truth as possible, by starting with what we know for sure—particulars. This would avoid confusions resulting from judging unknown or vague complexes that have not been traced to their sources. The other great advantage of nominalism, in the empiricist form in which it appears in Hume and Mach, is that the particulars are *sensed* particulars. This avoids reference to any external world beyond our perceptions, which, it is claimed, is unknowable and unnecessary for knowledge anyway.

We have surveyed one potential weakness of nominalism already, in Chapter 4. The motive for starting our claims to knowledge at the level of particulars is that we have immediate, objective, nontheoretical, and clear access to them in experience. But Copi and others suggest grave doubts about the existence of such an access. If there is no clear and objective "given," in experience, if all allegedly pure facts are actually context-bound, assumption-bound, and theory-bound facts, then the advantages of nominalism seem illusory. No "building blocks" for knowledge have been found.

A further difficulty with nominalism is the problem of specifying where relations come from. If, as Hume claimed, we experience two perceptions, say, an X and a Y, but not the relation between them, how do we come to assert any such relation? It won't do to answer merely that X is found in our experience to be always bigger than Y or more intense than Y or to the left of Y; for if that relation is really "found" or "discovered," we seem to be asserting that the relation is just as real and objective as the X and the Y. But that would be a denial of the nominalist thesis, because nominalism denies the reality of relations and of all other universals. Do we then invent or create the relation? If so, then why isn't such invention arbitrary, and why doesn't it vary widely from person to person? The nominalist might reply that our language and other conventions, molded and reinforced by survival needs and habits, have resulted in common inventions among people. But the critic of nominalism now urges that these are only stopgap explanations. Why do certain inventions and not others "work out" for our survival and comfort? Surely, just because certain relation assertions fit with the relations that are really "out there." More than human convention is involved when we learn that a hot stove has certain relations to tactile sensations. If our stipulated relations did not fit the world (of real relationships), we would get burned.

Finally, we should reiterate Wartofsky's point that for nominalists, counterfactual and subjunctive conditionals don't make sense. Nominalists believe that general statements are summaries of what actually happened or of what we expect to happen. Therefore, any "if . . . then" statement, where the "if" clause is known to be false or is merely theoretical, will have no reference. It will be as nonsensical as saying that "Blires are grite" when there is no such thing as a blire and no such quality as griteness. Those who believe that counterfactuals *do* make sense, and especially those who believe that they are important for science, will find nominalism seriously flawed in its position on this issue.[5]

Conceptualism in general, and Kant's position as one example of it in particular, is not likely to be supported initially by a student just beginning a study of ontology. It was proposed by Aristotle in ancient times, developed by Aquinas in the Middle Ages, and supported in a revised form by Kant in the eighteenth century, in each case as a response to the difficulties of the other positions. Yet once put forward, it does seem to have a plausibility that recommends it. Imagine two objects presented to you, say, a golf ball and a tennis ball. Now several relations obtain between these objects (objects that we may or may not grant to be simply bundles of perceptions). The tennis ball is larger, the golf ball is a lighter shade, and the two are a certain distance apart. Conceptualists urge that these relations exist just as surely and "objectively" as do the balls themselves. We do not, they will argue, see the two balls in isolation and then relate them. We see that they are related in size as soon as and just as surely as we see the objects themselves. We cannot ever experience these or other objects by themselves, alone, but rather always see them in some relation to other things. This seems to contradict the nominalist position. But it also seems clear that the relations exist only so long as the objects themselves exist. "Bigger than" is meaningless except in reference to existing objects. Conceptualists criticize realism for postulating an ideal world where relations exist without the need for any objects that are related. Conceptualists affirm the coexistent and interdependent reality of both objects, or perceptions, and the relations that obtain among them.

Some conceptualists believe that these relations and other universals are restricted to the world given in perception. Others hold that they belong to a world that exists independent of human perception. Aristotle, in the latter group, seems to have believed that they are real features of the world, to which the knowing mind conforms. Kant saw relations as impositions by the knowing mind that, together with the input of sense stimulations, give rise to perceptual experience.

We have said that realists claim to do justice to the meaning of law statements and other counterfactual and subjunctive conditionals. They deny that the other positions can do so. Conceptualists do have a way to interpret such conditionals, however, consistent with their position. Consider the following counterfactual:[6]

1. Had Earth and Venus collided in ancient times, they would have been destroyed.

Now if such a statement is to have meaning in conceptualist terms, it cannot literally refer to a relationship that existed even though the particular event did not exist. Relations and other universals, for the conceptualist, exist along with, but never apart from, particulars. Suppose, however, that we interpret the statement as a metaphorical way of saying the following:

2. Earth and Venus are known to behave like other large bodies, which leads us to predict that collisions of them will result in their destruction.

Statement 2 does not refer to a relationship between nonexistent events in the past. It is merely a prediction based on previous behavior. If statement 2 captures the meaning of statement 1, then conceptualists have done justice to counterfactuals consistently with their own position. Nominalists, too, if they can answer the other objections mentioned earlier, can avail themselves of the reformulation of counterfactuals into the form of statement 2.

Conceptualism has been attacked from both the realist and nominalist perspectives. Realists charge that it cannot account for the orderliness of the world. Objects are continually changing. The content of sense experience fluctuates continually. Universals, including relations and the laws that are their most general manifestation, cannot be as fleeting as particulars, say the realists. If they were, there would be no reason to expect the elegant, simple, and persistent order we find in the universe. By making the reality of relations, especially, dependent on the existence of particulars, conceptualists give them a chaotic (and unrealistic) character. Among the consequences of this mistake, realists say, is the twisted meaning conceptualists must give counterfactuals.

Nominalists challenge conceptualism's source of evidence for the existence of relations. They suspect that all relations have the same subjectivity and relativity as do obviously observer-dependent ones, such as "to the left of" or "sweeter than." Nominalists doubt the evidence for conceptualist theories such as Kant's, which hold that some relations are knowable a priori (logically prior to particular experiences). This battle concerning the possible existence of relations that are imposed by the knowing mind before experience is possible has continued in full force since Kant.[7]

We began this chapter by noting that the several alternative concepts of explanation discussed earlier all depend on commitments concerning the nature of reality. The commitment with which we have just dealt concerns the reality of universals. We now need to turn our attention to the second type of reality commitment mentioned earlier. Of what "stuff" is the real world composed? Are ideas real, is matter real, or is the world composed of both?

The Reality of Matter and Ideas

Materialism.   Materialism is among the oldest views on the nature of the universe. Fragments of texts indicate that as early as the sixth century B.C., the philosopher Thales was suggesting that the universe was composed of water in various stages of condensation. Somewhat later, Democritus and others defended atomism, the view that the universe consists of a collection of simple physical units. By Aristotle's time, it was becoming common to see the material universe as composed of air, earth, fire, and water. Modern materialists may hold either that the universe is composed of physical objects such as tables, chairs, and planets or that it is made up of some ultimately simple collection of subatomic particles. The "physical object" view is popular with those who insist that only the observable can qualify as real; the "simple particle" view is embraced by materialists who would allow unobservable bits of matter to qualify as real. These "theoretical entities," such as electrons, are not directly observable, but their postulation allows us to understand (and therefore explain?) observable events.

Exactly how does something qualify as material? Although criteria have varied somewhat from theory to theory, a material entity is usually thought to have (1) a definite location in space, (2) a definite extension (a border between inside and outside), and according to some definitions, (3) the capability of moving or being moved, and (4) impenetrability—or solidity. We will avoid demanding that it have mass, for that both ties the concept of matter to contemporary physics and raises problems with which we cannot deal in detail here.[8]

Materialism, as we are considering it, consists of two propositions. First, material things qualify as real; second, whatever is real must be material (that is, nothing that is *not* material can qualify as real).

How does a materialist deal with ideas, perceptions, and emotions, which are normally taken to be nonmaterial? Although answers have varied, two kinds of answers have predominated: the identity theory and epiphenomenalism. Both views share the basic materialist thesis that ideas, perceptions, and emotions are simply manifestations of underlying physical events. Now if only the physical occurrences are real, could we replace talk about perceptions and so forth with talk only about physical events? Could we, for example, do without the word *red* and substitute for it some expression about wavelengths, etc.? Identity theorists believe that such a translation is, at least in principle, possible. In contrast, epiphenomenalists argue that perceptual language has its own separate, though dependent, meaning; red, for example, would not exist without a certain wavelength of light, but at least some statements that refer to red could not be replaced by statements referring to wavelengths, without loss of meaning.[9]

What arguments can be given for materialism? To answer this question, we need to consider some of the most common reasons for believing that any given thing is real. The criteria that scientists and philosophers have used come surprisingly close to the ones we use every day in our ordinary experience. Among them are observability, simplicity, permanence, causality, and intelligibility. A brief explanation of these criteria will help us understand not only materialism but the other theories of reality as well.

1. *Observability.* Although some philosophers have been skeptical of observation because our sense experience is so variable and so capable of deceiving us, observability has continued to be one important criterion for locating the real, just as "seeing is believing" has long been important as a commonsense criterion. Materialists would argue that their theory is especially recommended by satisfying this criterion so well. Physical objects obviously are observable. Their tiny physical constituents may not be seen but are observable when taken in large clumps. In recent science, when subatomic particles are taken as the ultimate matter, some embarrassment is occasioned by the fact that these entities aren't even in principle observable. Materialists are likely to claim that such particles are at least "indirectly" observable because we can see the results of their interactions.

2. *Simplicity.* Simplicity is a deceptive criterion because there are so many types (compare, for example, the simplicity of the homunculus view of reproduction, which held that the male seed contains a miniature of the fully developed adult, with the simplicity claimed by Crick and Watson for the molecular structure of DNA). Nevertheless, simplicity is important in our everyday judgments about what is real because we confer reality status most readily on what is understandable, and often the simpler is the more understandable. A second reason for use of the simplicity criterion is that complexes come and go but the elements of those complexes remain, at least longer than the complexes do. These advantages of the simplicity criterion have been as important for scientists and philosophers as they have been for the layperson. Partly on these grounds, astronomers early saw the advantage of perfectly circular planetary motion, and in time it was partly simplicity (this time, of calculation) that led to the acceptance of Kepler's revision. Materialists have traditionally urged the great simplicity of reality on their theory, starting with the atomistic theories that saw all elements as identical. While a materialism that affirms the real existence of observable objects appears messy and complicated because of the

variety and complexity of such objects, materialists who confer reality on sub-atomic particles are in better shape. The continuing hope of such theorists is that we may be able to reduce the number of such particles in the interests of simplicity.

3. *Permanence*. As implied above, the everyday reliance on the criterion of simplicity is a derived one. Simplicity is important because of its relation to intelligibility and to permanence. We will discuss the intelligibility criterion below. Permanence has been an extremely important criterion from the beginnings of known ontological theories. However, in one sense it is itself derived. One reason we are interested in the permanence of the real is the possibility of knowledge it provides. No one values knowledge that may become outdated and false in the next instant. Knowledge that helps us survive in the future will have to be reliable, and therefore its object will have to "stay put." Above all, the object of reliable knowledge must not vanish, even if it changes form and relations often. On the grounds of permanence, materialists urge the reality of those ultimate particles that persist to form one complex object after another. The discovery of the convertibility of mass and energy obviously counts against the materialists view on this criterion. Materialists, though, can hope to retain the "ultimate particle" view by urging the move to a still more elementary particle level and the interpretation of energy as a quality of these ultimate elements.

4. *Causality*. We often take the real to be whatever accounts for appearances. We may respond to the charitable contributions of a public figure by saying, "What's *really* happening is that he is trying to curry public favor." Scientists may explain the illumination of a bulb by saying, "What's *really* happening is the movement of electrons through the wire." In these ways, our old friend causality, however often other words may be used in its stead, maintains an important place in the list of reality criteria. Clearly this criterion often leads us in a direction opposite that of observability. We have seen this collision of reality criteria manifested already, in the battle between phenomenalists and those who insist on a reality behind the appearances, which would account for them.

Materialists believe that their view fulfills the causality criterion as no other ontology can. They believe that brain state descriptions can account for our perceptions, and particle physics can account for physiological functions. For materialists, the world is ultimately matter in motion. Therefore, a description of that matter and the way it moves should account for all of the appearances we experience. Causal "accounting" does more than allow us to understand past and present phenomena. As we saw in Chapter 3, it promises justified predictiveness as well.

5. *Intelligibility*. Intelligibility is both fundamental (notice the partial dependence of all the other criteria on it) and elusive. Indeed, it may be as hard to characterize as Scriven (Chapter 5) believes understanding to be. The two terms have much in common. Following his suggestion, we might want to take *explaining* to mean "making intelligible." Our ordinary concept of intelligibility is also related to the concepts of familiarity, predictability, and reasonableness. Intelli-

gibility, as a criterion, makes us fit proposed realities into a coherent whole with the rest of what we accept as real. On these grounds, we exclude daydreams, chimeras, and other "claims" for the existence of things and events that on reflection we don't accept as real. Intelligibility also implies an order or arrangement that allows us to "fill in" the pieces we haven't experienced, just as a well-made jigsaw puzzle presents a picture and series of internal shapes that guide our search for the missing pieces. Materialists believe that a restriction of reality to material things alone fulfills the criterion of intelligibility. Appearances are presentations of whatever material realities they may be traced to and no others. Daydreams are manifestations of certain brain states and are real in that sense. They are "unreal" in the way they tempt us to believe certain matter is present when it really isn't. The search for the elementary motions of ultimate particles provides a single and easily understood order to nature that forms the internal intelligibility of apparently chaotic happenings in experience.

The case for materialism rests with the way it fulfills the preceding reality criteria: observability, simplicity, permanence, causality, and intelligibility. Similarly, those who disagree with materialism will argue for their alternative on the basis of reality criteria. Of course, it is possible for one to reject the list of criteria and therefore whatever position is argued on the basis of them. Productive disagreement about ontological positions has been possible, however, because at least general agreement about the criteria exists.

The sharpest disagreement with materialism has come from the position known as idealism, to which we will now turn.

Idealism. Idealists believe that reality is best described as like ideas, or perceptions, or processes. The long history of idealistic ontologies begins with philosophers before, and including, Plato in ancient Greece and continues with Alfred North Whitehead and Josiah Royce in our century.

Idealism is broad ground. It includes positions that disagree profoundly with one another. These positions share the view, however, that materialism fails to describe an observable, simple, permanent, causal, intelligible reality.

Those idealists who take the *observability* criterion most seriously are the phenomenalists. Following Berkeley (Chapter 4), they argue that matter isn't observable at all. When we see a table, we really only see a variety of qualities that we label as a clump. Therefore materialism, they say, fails this criterion completely.

As for *simplicity*, idealists charge that since the ancient hope for one kind of elementary particle, materialists have had to postulate an ever-increasing array of entities. The periodic table of the elements was bad enough; but now the proliferation of subatomic particles becomes downright embarrassing. By comparison, the processes of the universe, and the laws of nature that are their rules, appear quite simple. How much better to see the universe as made up of processes, of relationships, and/or of qualities than to see it as only a wide variety of material entities.

Modern physics, say the idealists, should have shattered any hope that materialism would yield a *permanent* ground for knowledge. Matter, if you take it as real, is convertible into energy, an entity quite different and one that does not fit the definition of matter we proposed earlier. Electrons seem to defy the demand that as material entities they have determinant motion and spatial location (though materialists may respond that the uncertainty principle represents our ignorance or the inadequacy of our model rather than a violation of their material status). Our current models provide no assurance that the ultimately simple elements of a matter-in-motion universe will match the classical definition of matter.

Concerning the *causality* criterion, idealists may argue that the materialist reference to particles does not provide causal knowledge. Just as seeing one billiard ball in motion succeeded by another billiard ball in motion does not mean having seen a causal relation between the two, so also observations of two subatomic particles (even if direct observation were possible) would not provide causal knowledge.

Finally, it might be charged (as a generalization of the point concerning causal relations) that materialism can admit of no real relations whatsoever—since relations between bits of matter are not themselves bits of matter. But to lack an account of relations would destroy the intelligibility of science. Relations, in the materialist theory, would become equivalent to certain states of matter, and this seems impossible.

If materialism is so unacceptable to idealists, what shape does their alternative account take, and how productive is their account in allowing us to understand scientific explanation? Two poles of idealism must be distinguished, though most idealist theories would fit somewhere in between. The difference between these two varieties of idealism concerns whether sense perceptions on the one hand or concepts (like number, cause, equation, fourteen, etc.) on the other are taken to be the real and the knowable. The variety that relies on sense perceptions as the real is phenomenalism, already cited as having its classical roots in the work of Berkeley. Phenomenalism's strength is its fulfillment of one traditionally important criterion of reality: observability. In Chapter 4 we noted that to rely on observability also seems to avoid the difficulty of making the real unknowable.

The weakness of phenomenalism is that if only discrete phenomena—fleeting sense impressions—are considered real, then two important sorts of things we normally think of as real in some sense are left unaccounted for: relations and minds, which themselves are surely more than bundles of perceptions (since, among other things, minds seem to be actors rather than being merely passive collections). In addition, other idealists argue that such concepts as mind, equality, cause, and so on, cannot be understood as mere abstractions from sense experience but, if they are understandable at all, must be understandable directly and in their own right.

The other pole of idealism attempts to retain the advantages of idealism without the foregoing weaknesses. It can be called conceptual idealism to recognize its focus on concepts, ideas, and the like, as the real. It may seem quite strange to imagine a world populated by concepts, laws, and other "spiritual" or "mentalistic" entities, without matter. This type of idealist, however, urges us to abandon our imagination (by which we are likely to mean "picturing") as the criterion of what is real. To rely on what we can imagine is to rely too much on the criterion of observability and correspondingly to underemphasize the importance of the other criteria for reality—especially permanence.

Plato, the greatest of the ancient idealists, insisted that if we could have knowledge at all, it couldn't be of sense impressions or of matter. Both of these are constantly changing and refiguring themselves. The musician learns that the numerical relationships of harmonies remain fixed and knowable however the various stringed instruments come and go. Only such permanent realities, of which particular strings and musical performances are instances, or copies, are capable of supporting an unchanging knowledge.

Conceptual idealism provides a plausible fulfillment of several of the reality criteria we have used. Among these, its particular strength is permanence, and its supporters would also give it high marks for intelligibility and simplicity. Its opponents, however, charge it with failing on two counts, causality and observability.[10] As for causality, conceptual idealists can claim to account well for formal causes, which we defined in Chapter 3 as the shape or form or pattern of events. For instance, the mathematical concepts of harmonies account for the musical tones produced. However, Aristotle criticized Plato for failing to use adequately efficient causality—the source of motion. The pushes and pulls of the world of our experience seemed to Aristotle to remain unintelligible when we focus on concepts alone. In general, idealists such as Plato interpreted the observable world of sense experience to be an unstable reflection[11] of the real world of ideas and not a criterion of the real in any clear sense. Critics of conceptual idealism charge that this emphasis means that our experience, which leads us to ask philosophical questions in the first place, is ignored rather than explained and interpreted.[12]

Dualism. The alternatives of materialism and idealism may have seemed extraordinarily harsh and extreme. Surely in our own lives we encounter both objects and ideas. Dualism is the theory that tries to combine the strengths of both materialism and idealism. It does so by affirming the existence of two kinds of realities: matter and ideas. Not only are we aware of these two kinds of entities, but we seem to experience their interaction, say the dualists. Illnesses, as well as a host of material changes for healthy people, may affect our ideas. Earthquakes cause fear; the sunset leads us to appreciation and wonder. And mounting evidence from the medical professions suggests the strong influence that ideas, hopes, and fears exert on our physical being.

Descartes provided us with the most famous classical dualistic metaphysics.[13] For him, as perhaps for us, dualism requires very little argument to make itself initially plausible. It seems to be simply an acknowledgment of the two kinds of reality we experience and that are reflected in our language. We are capable of expressing thoughts about both ideas and matter.

Some will protest that dualism is a fundamental violation of our criterion of simplicity. How messy to have a universe composed of two fundamentally different kinds of things! The great problems with dualism come not from this source, however, but from attempts to explain how two such unlike kinds of entities are related. That some account must be given of this relationship is obvious. Thoughts bear *some* relationship to brain states. Fears bear some relationship to illness. Numbers, formulae, laws, forces bear some relationship to the motion of particles of matter.

Descartes proposed the most obvious answer to the question of this relationship. He claimed, simply, that the world of ideas and the world of material objects, each with its own laws, interact. The question is, how? Survey the senses of cause reviewed in Chapter 3. Does a model seem appropriate? Do ideas move a billiard ball in the way another billiard ball does? Surely not. Does the law of gravity literally act to keep the earth in its orbit? Surely not. Hume suggested (Chapter 3) that what seems to be the most obvious interaction, where we choose for instance to raise our arm, comes, on analysis, to be bafflingly complex and perhaps utterly unintelligible. Scholars have puzzled over Descartes' weak answer to the question. He proposed that the mind and body interact through the pineal gland at the base of the brain. To the question of whether the pineal gland is itself ideal or material, there was no adequate answer. The ridicule of this proposal by subsequent thinkers has, however, been tempered by a recognition that no one else has proposed a better one. The simple and initially plausible position of interactionism, which would answer so many hard questions if only it were defensible, seems to hang on some answer to this puzzle of how matter and ideas interact.

The problems of interactionist dualism (especially explaining *how* the interaction takes place) have driven some dualists to adopt less initially plausible theories. Leibniz[14] proposed, for example, that the laws of the physical and the mental realms match like two clocks that are synchronized so that they appear to be one mechanism when they really are two separate ones. This doctrine is known as parallelism because it holds that occurrences in the mental world run parallel to those in the physical world.

The Double-Aspect Theory. Some theorists have been driven to either materialism or idealism, believing the problems of dualism are impossible to solve. Others, however, suspect that all of the alternatives mentioned so far are too narrow. When two observers stand on opposite sides of a mountain, they may describe it in such different terms that unless they are brought to accept it as the same mountain on other grounds, they might always believe themselves to have seen different objects. In one sense, of course, they *did* see different things, but in another and crucially important sense they were seeing different aspects of the same mountain. Could it be that the mental and the physical are simply two aspects in which a unitary reality presents itself—or, put more subjectively, two ways in which humans have come to see a unitary reality? Such a view is called the double-aspect theory.

The strength of such a position is that it refuses to acknowledge the confusions and dilemmas of the other alternatives. It does not, as we have just stated it, provide an alternative analysis that can show you why these difficulties are illusions. The double-aspect theory is not like other theories that have undisputed evidence in their favor. It is more like a proposal for a framework of thought, one that is cautious, tolerant, and skeptical of the ontological claims of the other theories. We may adhere to the double-aspect theory provisionally if we are unwilling to accept the weaknesses of materialism, idealism, or dualism. Yet acceptance of the double-aspect theory hardly answers the most pressing question: If reality has two or more aspects, how are they related?

Can we fruitfully argue about the relative merits of materialism, idealism, dualism, and double-aspect theories? No one of them seems provable with certainty. Yet each does have strengths and weaknesses that relate to the list of reality criteria: observability, simplicity, permanence, causality, and intelligibility. This relationship between positions and criteria makes argument and analysis worthwhile.

Reductionism

A critically important issue for the sciences, and for human life in general, has developed as a direct result of debates on the foregoing ontological positions. The issue is reductionism. As its name suggests, reductionism is the view that all descriptions of reality are reducible to one language about one type of being. Suppose nominalism and materialism of the identity variety are true, and true in such a way that all the other points of view are false. It would follow from this that an adequate science, and therefore adequate scientific explanation, ought to be expressible in one language about individual material things. It would be a clear, consistent language in which reference was made only to realities. Since these realities would be (1) particulars and (2) material entities, no reference to ideas, emotions, perceptions, or universals of any kind would be allowed unless such references were translatable into (that is, *reducible to*) statements that referred to the primary realities. Proponents of such a broad translation scheme are called ontological reductionists, and the appeal of their position is the clarity of such an "ultimate" language. Other reductionists who are not interested in advocating one ultimate set of ontological commitments may still urge that sociology is really translatable into psychology or that biology and chemistry are ultimately reducible to physics. In these reductionist suggestions, ontological claims are clearly not far in the background. One may advocate the reduction of biology to physics, for example, because of a belief that reality is ultimately simply a matter of particles in motion, and therefore that the behavior of organic complexes is understandable (and more clearly understandable) by describing the behavior of their constituent parts. By contrast, it is clear that double-aspect theorists are likely to criticize attempts at reduction. To describe one side of a mountain in terms of another is not likely to promote understanding!

Reductionism is clearly an important issue for our account of explanation. Like the battles described in Chapter 3 concerning the translation of one kind of causal explanation into another, debates on the reduction of one discipline's explanations to those of another concern the direction of teaching and research in a whole field.

Countering the appeal to simplicity and clarity, opponents of reductionism frequently claim that essential areas of understanding will be missed, or simply ignored, if reduction is attempted on any broad scale. We will consider these arguments in some detail in Chapter 9. At this point, it is important for us simply to notice that the ontological positions we adopt will directly influence our view of the possibility and desirability of reduction.

Realism/Nominalism and Materialism/Idealism

We have explored the range of positions on the reality of universals and the reality of matter and ideas separately. Our search for a general view of explanation demands that we also see how combinations of commitments fit together. How are views on the two major ontological questions of this chapter related to each other?

The two columns below summarize the positions we have discussed concerning the two main ontological questions.

The Reality of Universals	*The Reality of Matter and Ideas*
1. Realism	1. Materialism
2. Nominalism	a. Identity thesis
3. Conceptualism	b. Epiphenomenalism
	2. Idealism
	a. Phenomenalism
	b. Conceptual idealism
	3. Dualism
	a. Interactionism
	b. Parallelism
	4. Double-aspect theory

Which combinations of these two lists provide the most likely alternative accounts of the nature of reality? Which combinations are most unlikely?

First, consider two likely combinations. Nominalists will most often be found holding either the particle form of materialism, or phenomenalism. The crucial common thrust of all these positions is emphasis on the particular and denial of the reality of universals that, on these accounts, are merely abstractions—useful perhaps, but not real. Nominalism asserts the primary reality of particulars of some sort, while phenomenalism and particle materialism provide alternative accounts of what these particulars are like.

Several combinations are unlikely. Realists, believing that universals are real, are not going to subscribe to materialism of any variety. This is so because material things are particular entities, whereas universals are conceptual, or ideal, and not material. One could not affirm the reality of universals and also uphold materialism, which denies the reality of nonmaterial things.[15]

As you begin to try combinations of commitments that make sense to you, it is a good exercise to throw potentially compatible hypotheses together and then develop a sense of what sort of universe is specified by those commitments.

Consider, for example, what view of the universe issues from a combination of nominalism and dualism. It will consist of a collection of particular physical units and a collection of particular mental entities, either sense impressions or concepts. The central problem in describing it will be to account for the interaction (either real or apparent) of the physical and the mental. Relations and the laws that summarize their universal forms will be abstract summaries of experience imposed by the mind in an effort to classify and simplify. Counterfactual assertions will likely be held to be meaningless or, at best, metaphorical.

Ontology and Explanation

We began this chapter by recognizing that our choice of accounts of scientific explanation will depend on the way we view reality and our ability to know it. Now let us return to the views of explanation we have considered to see what ontological commitments are demanded by them.

This brief review of some themes of earlier chapters will constitute "practice problems" in applying our knowledge of ontological positions discussed in this chapter.

The causal view of scientific explanation (Chapter 3) affirms the reality of causal relationships. Since these relations are universals, causalists will therefore need to defend either realism or conceptualism. Furthermore, their need to affirm the reality of relations seems to exclude materialism and phenomenalism.

Hume himself seems closest to nominalism because he denied that we can have rational knowledge of relationships in nature. Some commentators, though, have argued that his insistence on the impositions of relations on phenomena, by the imagination, draws him close to a rudely carved conceptualism.

He also seems at times to have hoped for a thoroughgoing materialistic account of nature and knowing. Such a hope was rendered impossible of fulfillment by his own theory, however. The most prevailing conclusion from his theory is certainly that any ontological commitments beyond a straightforward phenomenalism are groundless and only the source of fruitless disputes.

We saw in Toulmin's and Russell's positions on causality (Chapter 3), a rejection of ontological causal claims but a continued reliance on the future uniformity of nature. Now this trust in the uniformity of nature makes necessary some ontology that renders the uniformity intelligible. Toulmin seems to acknowledge the need for commitment on this issue and recognizes the uniformity hypothesis as one of these assumptions. He argues that science generally involves clusters of assumptions combined in a framework of commitments.[16]

Kant's answer to Hume's criticism of causal knowledge claims led to one classical form of conceptualism, as mentioned by Wartofsky. While Kant denied the possibility of ontological knowledge of a reality beyond possible perception, the reality of human experience was, for Kant, phenomenal. The real, for science, is the perceptual world of rational creatures, complete with the relations and other concepts that inform it. This is not quite phenomenalism, for "percepts without concepts are blind."[17] Kant's idealist doctrine therefore acknowledges the existence in experience of both concepts and sense perceptions, united indissolubly.

Positivists are the modern proponents of nominalism. The heart of their program involves the denial of the reality of "abstract" universals. Positivists have sometimes embraced materialism, or dualism, or, like Mach, phenomenalism. True to their antimetaphysical spirit, they have often tried to avoid these commitments, however, focusing their work on the logic of scientific language or the methodology of science. In this tradition we find the modern instrumentalists, so named because of their view that theories, rather than making truth claims about the reality of the entities they mention, are actually simply instruments for describing and predicting sense experiences. Positivists and instrumentalists still find it exceedingly difficult to account for a faith in the future regularity of nature without making metaphysical commitments.

If we see the covering law model as the strongest account or part of the strongest account of scientific explanation, to what ontological theories are we committed? To answer this question, we must remember that the model only claims to expose the logical structure of explanation. Hempel and Oppenheim do not tell us where the crucial first premise—the lawlike statement—in the deduction comes from. Therefore, the model itself does not commit its proponents to any particular theory on the reality of universals. Universals may be mere descriptive summaries of past sense experience, and this would make the model consistent with nominalism. If these first premises are taken to be statements of real relations, the model fits into a conceptualist or realist scheme.

The covering law model can, in like fashion, be made consistent with any of the materialist/idealist alternatives. The model is particularly attractive to reductionists (either idealist or materialist), who may see it as the logical path to the construction of one unified scientific language, ultimately clear and ultimately simple.

Common Ground and Alternatives

Scientists make judgments about the nature of reality. Our choice of the most adequate *explanations* will reflect other, broader choices of *ontological theories*. We have seen that these choices concern, especially, two questions: first, whether universals or only particulars are real, and second, whether matter, or ideas, or both are real. Among other issues, reductionism depends heavily on one's view of these alternative ontologies. Finally, we have been able to identify the classical theories of explanation by their dependence on various ontological commitments.

At the beginning of this chapter, we suggested that, for all their differences, the approaches to scientific explanation we have considered share common ground. Specifically, they see explanation as seeking a reality that, however humanly conditioned it may be, is largely independent of our will. Even if reality consists of our perceptions, those perceptions are presented to us rather than being created by us. We questioned whether any alternative to this common view is possible.

The alternative with which we will deal in Chapter 7 asks us to consider human willing as central to the process of scientific explanation. Therefore, in its starkest form, this alternative stands in marked contrast to the positions so far considered.

Supplementary Readings

Taylor, Richard. *Metaphysics.* Englewood Cliffs, N.J.: Prentice-Hall, 1963.

This introduction to some of the problems of metaphysics does not assume background knowledge in philosophy. A number of other volumes in the Prentice-Hall Foundations of Philosophy Series are also likely to prove useful.

Copleston, Frederick C. *Medieval Philosophy.* New York: Harper, 1961.

This introduction to the philosophical systems and debates of the Middle Ages puts in a broader context the spectrum of opinion on the question of the reality of universals.

Notes

1. ". . . the species of cause termed final . . . finds no useful employment in physical (or natural) things; for it does not appear to me that I can without temerity seek to investigate the (inscrutible) ends of God." From meditation IV, *Meditations*, Vol. 1, *The Philosophical Works of Descartes*, trans. E. S. Haldane and G. R. T. Ross (London: Cambridge University Press, 1911), p. 173.

2. Marx Wartofsky, *Conceptual Foundations of Scientific Thought* (New York: Macmillan, 1968), pp. 249–58. It is especially good for our purposes that this discussion of the reality of universals comes from a philosopher of science. Using scientific laws as an example of a universal keeps this discussion of philosophy close to science and scientific explanation. It also provides some basic clarity on scientific laws.

3. James D. Watson, *The Double Helix* (New York: Atheneum, 1968).

4. We here have referred to nominalism "in its empiricist form." What would a nonempirical nominalism be like? By definition, such a position would hold that only particulars are real, but that some particulars may be known by means other than experience.

5. For a further discussion of the importance of counterfactuals, see Nelson Goodman, *Fact, Fiction, and Forecast* (Cambridge, Mass: Harvard University Press, 1955).

6. Cf. Chapter 2; note that the falsity of the premise (the "if" clause) makes the statement a counterfactual.

7. One recent debate relevant to this question concerns Jean Piaget's psychological model of human intellectual development, which seems to some

critics to employ reference to a priori mental structures. Another debate concerns the work of Noam Chomsky in linguistics, which uses the notion of a universal cognitive structure underlying all languages. The behaviorist critics of Piaget (e.g., Susan Isaacs) and of Chomsky (e.g., B. F. Skinner) uphold the nominalist side of these recent arguments.

8. For example, are massless "particles" like the neutrino material?

9. Epiphenomenalism got its name because nonmaterial events are taken to be surface (epi) phenomena produced by an underlying material reality.

10. Aristotle criticized Plato's theory on a number of grounds not limited to those mentioned. Plato scholarship becomes complicated and fascinating at this point, however. Some of Plato's work indicates that he must have been aware of these criticisms and saw his theory in more complex terms than the summary we have had to give would suggest. For instance, in the *Parmenides*, Plato had a critic argue against the apparent argument of the *Republic*, and apparently win.

11. The "reflection" metaphor is used intentionally. Note Plato's "Allegory of the Cave" in the *Republic*. One readable edition is *The Republic of Plato*, trans. F. M. Cornford (New York: Oxford University Press, 1945).

    The initial clarity of the apparent positions of Plato and Aristotle, the subtlety of the questions as one begins to study them in detail, and the relevance and power of the two positions as frameworks within which later thinking has taken place, all combine to make the study of these thinkers especially rewarding for students of philosophy at all levels.

12. The challenge for later idealists, including Hegel and more recently F. H. Bradley, has been to face the charge that they ignore experience and to argue that sense experience either is interpretable only on their terms or, as mere appearance, is unworthy of philosophical study.

13. René Descartes, *Meditations on First Philosophy*, in *The Philosophical Works of Descartes*, trans. E. S. Haldane and G. R. T. Ross (London: Cambridge University Press, 1911).

14. Leibniz, *Monodology* (1714), *Discourse on Metaphysics* (1686), *New Essays on the Understanding* (1704), in *Leibniz, Selections*, ed. Philip Wiener (New York: Scribner's, 1951).

15. For those with some background in the philosophy of science we should offer the reminder that we are here talking of philosophical realism, not scientific realism. We will enlarge on this distinction in Chapter 8.

16. Stephen Toulmin, *The Philosophy of Science* (London: Hutchinson, 1953).

17. Immanuel Kant, *Critique of Pure Reason* (1781), trans. Norman Kemp Smith (New York: St. Martin's, 1961).

/7/ THE PRAGMATIST ANSWER

Proposal: An Adequate Scientific Explanation Is One That Is Useful

"Rust Is a Reality" reads the rustproofing ad, and we ruefully nod our heads in agreement as we consider the fenders and doors of a car that looked so beautiful and shiny not so long ago. Whether we like it or not, whether we can explain the process of corrosion scientifically or not, we do have to admit that nature runs its course and salt accelerates the process. Trying to understand the rusting of steel seems very far away from the difficult questions of philosophy we have been considering. Whether you consider rust to be material or ideal, rust is a reality you must deal with. Perhaps your local rustproofer is not the person who can banish rust forever, but someone has to work on problems like that. Long arguments about the form of scientific explanation are not very satisfying to someone who expects science to solve practical problems. Perhaps the best scientific theories are those that offer practical advantages, and the most scientific explanation is the one that helps us best face the reality of rust.

Positivists, causalists, and others whom we have discussed in chapters 3 through 5 will react with firm disapproval to the demand that the central purpose of science be the solution of practical problems. Scientists, they will insist, are not engineers or technicians. Scientists are not primarily concerned with solving human practical problems but rather are engaged in a search for the truth about our universe. Knowing more about the world we live in will help us solve problems but not the other way around. Scientists don't work on the rust problem in order to determine the goals and methods of science. We should allow ourselves to ask, "What good is it?" only after we have learned the truth about it.

Some philosophers of science, however, have argued persistently that considerations of values, of payoffs and usefulness are in fact what do finally lead us to choose some theories over others. Such philosophers provide us with an alternative to the "common ground" of the theories discussed earlier. This alternative, most clearly stated in the school of thought called pragmatism, holds that "the true" means nothing else but the useful. In this view, it is not the case that what is true just happens to turn out to be useful. It could turn no other way because the two are related by definition. Pragmatists could point out that we should have seen this result emerging from the development of the argument in the preceding chapters. We noticed that the hope for a given in human experience or thought is an element in several theories we have considered. Causalists, responding to Hume's criticism, may urge that causal relations are either directly experienced (as in the criticism of Hume's "atomism") or are direct correspondences in which thoughts mirror reality. Positivists, seeking a guaranteed objective foundation in sense experience, are likely to affirm such an immediate given in our sense impressions. Yet each appeal to a given where the facts stand by themselves, recognizable without recourse to frameworks of assumptions, came in for serious (though perhaps not fatal) criticism. This criticism turned our attention to choices of explanation frameworks, among them the covering law model. Arguments for this, or Scriven's, or any other theory of explanation could hardly be decided by an appeal to the bare facts of experience. After all, these theories provide frameworks within which these facts themselves are interpreted and assessed for significance. But if the facts themselves do not provide us with a choice among these models, what does decide the issue? "Usefulness," say the pragmatists. "That model of explanation is best that helps us answer the questions that are *important* to us! The search for a picture of adequate scientific explanations leads us to the question of what is ultimately useful."

Pragmatism found its classical expression in the late nineteenth and early twentieth centuries in the work of three American philosophers: William James, C. S. Peirce, and John Dewey. Though these thinkers did not agree on all points, they shared the general view that truth is more like a process, a human activity, than like a static correspondence with an unchanging reality. The pragmatists acknowledged their debts to other theories, particularly to the empiricist tradition following from Hume; but they brought a new emphasis to the ties between knowledge and action, between facts and values.

In one way, James is a good representative of the pragmatists; in other ways he is not. The clarity and force of his writing did much to popularize pragmatism; the starkness of some of his argument made clear what was at stake. Since he did not develop the position with the subtlety and power of Peirce, however, his presentation should not be seen as *the* pragmatist position. We have included, in the following selection, excerpts from two of his most famous essays. Although they are highly rhetorical at points, they are unparalleled in stating the pragmatist thesis crisply and persuasively.[1]

WHAT PRAGMATISM MEANS

The pragmatic method is primarily a method of settling metaphysical disputes that otherwise might be interminable. Is the world one or many?—fated or free?—material or spiritual?—here are notions either of which may or may not hold good of the world; and disputes over such notions are unending. The pragmatic method in such cases is to try to interpret each notion by tracing its respective practical consequences. What difference would it practically make to any one if this notion rather than that notion were true? If no practical difference whatever can be traced, then the alternatives mean practically the same thing, and all dispute is idle. Whenever a dispute is serious, we ought to be able to show some practical difference that must follow from one side or the other's being right.

A glance at the history of the idea will show you still better what pragmatism means. The term is derived from the same Greek word πράγμα, meaning action, from which our words "practice" and "practical" come. It was first introduced into philosophy by Mr. Charles Peirce in 1878. In an article entitled "How to Make Our Ideas Clear," in the Popular Science Monthly for January of that year, Mr. Peirce, after pointing out that our beliefs are really rules for action, said that, to develop a thought's meaning, we need only determine what conduct it is fitted to produce: that conduct is for us its sole significance. And the tangible fact at the root of all our thought-distinctions, however subtle, is that there is no one of them so fine as to consist in anything but a possible difference of practice. To attain perfect clearness in our thoughts of an object, then, we need only consider what conceivable effects of a practical kind the object may involve—what sensations we are to expect from it, and what reactions we must prepare. Our conception of these effects, whether immediate or remote, is then for us the whole of our conception of the object, so far as that conception has positive significance at all.

This is the principle of Peirce, the principle of pragmatism. It lay entirely unnoticed by any one for twenty years, until I, in an address before Professor Howison's Philosophical Union at the University of California, brought it forward again and made a special application of it to religion. By that date (1898) the times seemed ripe for its reception. The word "pragmatism" spread, and at present it fairly spots the pages of the philosophic journals. On all hands we find the "pragmatic movement" spoken of, sometimes with respect, sometimes with contumely, seldom with clear understanding. It is evident that the term applies itself conveniently to a number of tendencies that hitherto have lacked a collective name, and that it has "come to stay."

To take in the importance of Peirce's principle, one must get accustomed to applying it to concrete cases. I found a few years ago that Ostwald, the illustrious Leipzig chemist, had been making perfectly distinct use of the principle of pragmatism in his lectures on the philosophy of science, though he had not called it by that name.

"All realities influence our practice," he wrote me, "and that influence is their meaning for us. I am accustomed to put questions to my classes in this way: In what respects would the world be different if this alternative or that were true? If I can find nothing that would become different, then the alternative has no sense."

That is, the rival views mean practically the same thing, and meaning, other than practical, there is for us none. Ostwald in a published lecture gives this example of what he means. Chemists have long wrangled over the inner constitution of certain bodies called "tautomerous." Their properties seemed equally consistent with the notion that an instable hydrogen atom oscillates inside of them, or that they are instable mixtures of two bodies. Controversies raged, but never was decided. "It would never have begun," says Ostwald, "if the combatants had asked themselves what particular experimental fact could have been made different by one or the other view being correct. For it would then have appeared that no difference of fact could possibly ensue; and the quarrel was as unreal as if, theorizing in primitive times about the raising of dough by yeast, one party should have invoked a 'brownie,' while another insisted on an 'elf' as the true cause of the phenomenon."*

It is astonishing to see how many philosophical disputes collapse into insignificance the moment you subject them to this simple test of tracing a concrete consequence. There can be no difference anywhere that doesn't make a difference elsewhere—no difference in abstract truth that doesn't express itself in a difference in concrete fact and in conduct consequent upon that fact, imposed on somebody, somehow, somewhere, and somewhen. The whole function of philosophy ought to be to find out what definite difference it will make to you and me, at definite instants of our life, if this world-formula or that world-formula be the true one.

There is absolutely nothing new in the pragmatic method. Socrates was an adept at it. Aristotle used it methodically. Locke, Berkeley, and Hume made momentous contributions to truth by its means. Shadworth Hodgson keeps insisting that realities are only what they are "known as." But these forerunners of pragmatism used it in fragments: they were preluders only. Not until in our time has it generalized itself, become conscious of a universal mission, pretended to a conquering destiny. I believe in that destiny, and I hope I may end by inspiring you with my belief.

Pragmatism represents a perfectly familiar attitude in philosophy, the empiricist attitude, but it represents it, as it seems to me, both in a more radical and in a less objectionable form than it has ever yet assumed. A pragmatist turns his back resolutely and once for all upon a lot of inveterate habits dear to professional philosophers. He turns away from abstraction and insufficiency, from verbal solutions, from bad a priori reasons, from fixed principles, closed systems, and pretended absolutes and origins. He turns towards concreteness and adequacy, towards facts, towards action, and towards power. That means the empiricist temper regnant and the rationalist temper sincerely given up. It means the open air and possibilities of nature, as against dogma, artificiality, and the pretense of finality in truth.

At the same time it does not stand for any special results. It is a method only. But the general triumph of that method would mean an enormous change in what I called in my last lecture the "temperament" of philosophy. Teachers of the ultra-rationalistic type would be frozen out, much as the courtier type is frozen out in republics, as the ultramontane type of priest is frozen out in protestant lands. Science and metaphysics would come much nearer together, would in fact work absolutely hand in hand.

Metaphysics has usually followed a very primitive kind of quest. You know how men have always hankered after unlawful magic, and you know what a great part in magic words have always played. If you have his name, or the formula of incantation that binds him, you can control the spirit, genie, afrite, or whatever the power may be.

*"Theorie und Prazis," Zeitschrift des Oesterreichischen Ingenieur-u. Architecten-Vereins, 1905, Nr. 4 u. 6. I find a still more radical pragmatism than Ostwald's in an address by Professor W. S. Franklin: "I think that the sickliest notion of physics, even if a student gets it, is that it is 'the science of masses, molecules, and the ether.' And I think that the healthiest notion, even if a student does not wholly get it, is that physics is the science of the ways of taking hold of bodies and pushing them!" (Science, January 2, 1903).

Solomon knew the names of all the spirits, and having their names, he held them subject to his will. So the universe has always appeared to the natural mind as a kind of enigma, of which the key must be sought in the shape of some illuminating or power-bringing word or name. That word names the universe's principle, and to possess it is after a fashion to possess the universe itself. "God," "Matter," "Reason," "the Absolute," "Energy," are so many solving names. You can rest when you have them. You are at the end of your metaphysical quest.

But if you follow the pragmatic method, you cannot look on any such word as closing your quest. You must bring out of each word its practical cash-value, set it at work within the stream of your experience. It appears less as a solution, then, than as a program for more work, and more particularly as an indication of the ways in which existing realities may be changed.

Theories thus become instruments, not answers to enigmas, in which we can rest. We don't lie back upon them, we move forward, and, on occasion, make nature over again by their aid. Pragmatism unstiffens all our theories, limbers them up and sets each one at work. Being nothing essentially new, it harmonizes with many ancient philosophic tendencies. It agrees with nominalism, for instance, in always appealing to particulars; with utilitarianism in emphasizing practical aspects; with positivism in its disdain for verbal solutions, useless questions and metaphysical abstractions.

All these, you see, are anti-intellectualist tendencies. Against rationalism as a pretension and a method pragmatism is fully armed and militant. But, at the outset, at least, it stands for no particular results. It has no dogmas, and no doctrines save its method. As the young Italian pragmatist Papini has well said, it lies in the midst of our theories, like a corridor in a hotel. Innumerable chambers open out of it. In one you may find a man writing an atheistic volume; in the next some one on his knees praying for faith and strength; in a third a chemist investigating a body's properties. In a fourth a system of idealistic metaphysics is being excogitated; in a fifth the impossibility of metaphysics is being shown. But they all own the corridor, and all must pass through it if they want a practicable way of getting into or out of their respective rooms.

No particular results then, so far, but only an attitude of orientation, is what the pragmatic method means. The attitude of looking away from first things, principles, "categories," supposed necessities; and of looking towards last things, fruits, consequences, facts. . . .

One of the most successfully cultivated branches of philosophy in our time is what is called inductive logic, the study of the conditions under which our sciences have evolved. Writers on this subject have begun to show a singular unanimity as to what the laws of nature and elements of fact mean, when formulated by mathematicians, physicists, and chemists. When the first mathematical, logical, and natural uniformities, the first laws, were discovered, men were so carried away by the clearness, beauty, and simplification that resulted, that they believed themselves to have deciphered authentically the eternal thoughts of the Almighty. His mind also thundered and reverberated in syllogisms. He also thought in conic sections, squares and roots and ratios, and geometrized like Euclid. He made Kepler's laws for the planets to follow; he made velocity increase proportionally to the time in falling bodies; he made the law of the sines for light to obey when refracted; he established the classes, orders, families, and genera of plants and animals, and fixed the distances between them. He thought *the archetypes of all things, and devised their variations; and when we rediscover any one of these his wondrous institutions, we seize his mind in its very literal intention.*

But as the sciences have developed further, the notion has gained ground that most, perhaps all, of our laws are only approximations. The laws themselves, moreover, have grown so numerous that there is no counting them; and so many rival formulations are proposed in all the branches of science that investigators have become accustomed to the notion that no theory is absolutely a transcript of reality, but that any one of them may from some point of view be useful. Their great use is to summarize old facts and to lead to new ones. They are only a man-made language, a conceptual shorthand, as some one calls them, in which we write our reports of nature; and languages, as is well known, tolerate much choice of expression and many dialects.

Thus human arbitrariness has driven divine necessity from scientific logic. If I mention the names of Sigwart, Mach, Ostwald, Pearson, Milhaud, Poincare, Duhem, Ruyssen, those of you who are students will easily identify the tendency I speak of, and will think of additional names.

Riding now on the front of this wave of scientific logic Messrs. Schiller and Dewey appear with their pragmatistic account of what truth everywhere signifies. Everywhere, these teachers say, "truth" in our ideas and beliefs means the same thing that it means in science. It means, they say, nothing but this, that ideas (which themselves are but parts of our experience) become true just in so far as they help us to get into satisfactory relation with other parts of our experience, to summarize them and get about among them by conceptual shortcuts instead of following the interminable succession of particular phenomena. Any idea upon which we can ride, so to speak; any idea that will carry us prosperously from any one part of our experience to any other part, linking things satisfactorily, working securely, simplifying, saving labor; is true for just so much, true in so far forth, true instrumentally. This is the "instrumental" view of truth taught so successfully at Chicago, the view that truth in our ideas means their power to "work," promulgated so brilliantly at Oxford.

Messrs. Dewey, Schiller, and their allies, in reaching this general conception of all truth, have only followed the example of geologists, biologists, and philologists. In the establishment of these other sciences, the successful stroke was always to take some simple process actually observable in operation—as denudation by weather, say, or variation from parental type, or change of dialect by incorporation of new words and pronunciations—and then to generalize it, making it apply to all times, and produce great results by summating its effects through the ages.

The observable process which Schiller and Dewey particularly singled out for generalization is the familiar one by which any individual settles into new opinions. The process here is always the same. The individual has a stock of old opinions already, but he meets a new experience that puts them to a strain. Somebody contradicts them; or in a reflective moment he discovers that they contradict each other; or he hears of facts with which they are incompatible; or desires arise in him which they cease to satisfy. The result is an inward trouble to which his mind till then had been a stranger, and from which he seeks to escape by modifying his previous mass of opinions. He saves as much of it as he can, for in this matter of belief we are all extreme conservatives. So he tries to change first this opinion, and then that (for they resist change very variously), until at last some new idea comes up which he can graft upon the ancient stock with a minimum of disturbance of the latter, some idea that mediates between the stock and the new experience and runs them into one another most felicitously and expediently.

This new idea is then adopted as the true one. It preserves the older stock of truths with a minimum of modification, stretching them just enough to make them admit the novelty, but conceiving that in ways, as familiar as the case leaves possible. An outré explanation, violating all our preconceptions, would never pass for a true account of a novelty. We should scratch round industriously till we found something less eccentric. The most violent revolutions in an individual's beliefs leave most of his old order standing. Time and space, cause and effect, nature and history, and one's own biography remain untouched. New truth is always a go-between, a smoother-over of transitions. It marries old opinion to new fact so as ever to show a minimum of jolt, a maximum of continuity. We hold a theory true just in proportion to its success in solving this "problem of maxima and minima." But success in solving this problem is eminently a matter of approximation. We say this theory solves it on the whole more satisfactorily than that theory; but that means more satisfactorily to ourselves, and individuals will emphasize their points of satisfaction differently. To a certain degree, therefore, everything here is plastic.

The point I now urge you to observe particularly is the part played by the older truths. Failure to take account of it is the source of much of the unjust criticism leveled against pragmatism. Their influence is absolutely controlling. Loyalty to them is the first principle—in most cases it is the only principle; for by far the most usual way of handling phenomena so novel that they would make for a serious rearrangement of our preconception is to ignore them altogether, or to abuse those who bear witness for them.

You doubtless wish examples of this process of truth's growth, and the only trouble is their superabundance. The simplest case of new truth is of course the mere numerical addition of new kinds of facts, or of new single facts of old kinds, to our experience—an addition that involves no alteration in the old beliefs. Day follows day, and its contents are simply added. The new contents themselves are not true, they simply come and are. Truth is what we say about them, and when we say that they have come, truth is satisfied by the plain additive formula.

But often the day's contents oblige a rearrangement. If I should now utter piercing shrieks and act like a maniac on this platform, it would make many of you revise your ideas as to the probable worth of my philosophy. "Radium" came the other day as part of the day's content, and seemed for a moment to contradict our ideas of the whole order of nature, that order having come to be identified with what is called the conservation of energy. The mere sight of radium paying heat away indefinitely out of its own pocket seemed to violate that conservation. What to think? If the radiations from it were nothing but an escape of unsuspected "potential" energy, pre-existent inside of the atoms, the principle of conservation would be saved. The discovery of "helium" as the radiation's outcome, opened a way to this belief. So Ramsay's view is generally held to be true, because, although it extends our old ideas of energy, it causes a minimum of alteration in their nature.

I need not multiply instances. A new opinion counts as "true" just in proportion as it gratifies the individual's desire to assimilate the novel in his experience to his beliefs in stock. It must both lean on old truth and grasp new fact; and its success (as I said a moment ago) in doing this, is a matter for the individual's appreciation. When old truth grows, then, by new truth's addition, it is for subjective reasons. We are in the process and obey the reasons. That new idea is truest which performs most felicitously its function of satisfying our double urgency. It makes itself true, gets itself classed as true, by the way it works; grafting itself then upon the ancient body of truth, which thus grows much as a tree grows by the activity of a new layer of cambium.

Now Dewey and Schiller proceed to generalize this observation and to apply it to the most ancient parts of truth. They also once were plastic. They also were called true for human reasons. They also mediated between still earlier truths and what in those days were novel observations. Purely objective truth, truth in whose establishment the function of giving human satisfaction in marrying previous parts of experience with newer parts played no role whatever, is nowhere to be found. The reasons why we call things true is the reason why they are true, for "to be true" means only to perform this marriage-function.

The trail of the human serpent is thus over everything. Truth independent; truth that we find merely; truth no longer malleable to human need; truth incorrigible, in a word; such truth exists indeed superabundantly—or is supposed to exist by rationalistically minded thinkers; but then it means only the dead heart of the living tree, and its being there means only that truth also has its paleontology, and its "prescription," and may grow stiff with years of veteran service and petrified in men's regard by sheer antiquity. But how plastic even the oldest truths nevertheless really are has been vividly shown in our day by the transformation of logical and mathematical ideas, a transformation which seems even to be invading physics. The ancient formulas are reinterpreted as special expressions of much wider principles, principles that our ancestors never got a glimpse of in their present shape and formulation.

I am well aware how odd it must seem to some of you to hear me say that an idea is "true" so long as to believe it is profitable to our lives. That it is good, for as much as it profits, you will gladly admit. If what we do by its aid is good, you will allow the idea itself to be good in so far forth, for we are the better for possessing it. But is it not a strange misuse of the word "truth," you will say, to call ideas also "true" for this reason?

To answer this difficulty fully is impossible at this stage of my account. You touch here upon the very central point of Messrs. Schiller's, Dewey's, and my own doctrine of truth, which I cannot discuss with detail until my sixth lecture. Let me now say only this, that truth is one species of good, and not, as is usually supposed, a category distinct from good, and coordinate with it. The true is the name of whatever proves itself to be good in the way of belief, and good, too, for definite, assignable reasons. Surely you must admit this, that if there were no good for life in true ideas, or if the knowledge of them were positively disadvantageous and false ideas the only useful ones, then the current notion that truth is divine and precious, and its pursuit a duty, could never have grown up or become a dogma. In a world like that, our duty would be to shun truth, rather. But in this world, just as certain foods are not only agreeable to our taste, but good for our teeth, our stomach, and our tissues; so certain ideas are not only agreeable to think about, or agreeable as supporting other ideas that we are fond of, but they are also helpful in life's practical struggles. If there be any life that it is really better we should lead, and if there be any idea which, if believed in, would help us to lead that life, then it would be really better for us to believe in that idea, unless, indeed, belief in it incidentally clashed with other greater vital benefits.

"What would be better for us to believe!" This sounds very like a definition of truth. It comes very near to saying "what we ought to believe"; and in that definition none of you would find any oddity. Ought we ever not to believe what it is better for us to believe? And can we then keep the notion of what is better for us, and what is true for us, permanently apart?

Pragmatism says no, and I fully agree with her. Probably you also agree, so far as the abstract statement goes, but with a suspicion that if we practically did believe everything that made for good in our own personal lives, we should be found indulging in all kinds of fancies about this world's affairs, and all kinds of sentimental superstitions about a world hereafter. Your suspicion here is undoubtedly well founded, and it is evident that something happens when you pass from the abstract to the concrete that complicates the situation.

I said just now that what is better for us to believe is true unless the belief incidentally clashes with some other vital benefit. Now in real life what vital benefits is any particular belief of ours most liable to clash with? What indeed except the vital benefits yielded by other beliefs when these prove incompatible with the first ones? In other words, the greatest enemy of any one of our truths may be the rest of our truths. Truths have once for all this desperate instinct of self-preservation and of desire to extinguish whatever contradicts them. . . .

PRAGMATISM'S CONCEPTION OF TRUTH

Let me begin by reminding you of the fact that the possession of true thoughts means everywhere the possession of invaluable instruments of action; and that our duty to gain truth, so far from being a blank command from out of the blue, or a "stunt" self-imposed by our intellect, can account for itself by excellent practical reasons.

The importance to human life of having true beliefs about matters of fact is a thing too notorious. We live in a world of realities that can be infinitely useful or infinitely harmful. Ideas that tell us which of them to expect count as the true ideas in all this primary sphere of verification, and the pursuit of such ideas is a primary human duty. The possession of truth, so far from being here an end in itself, is only a preliminary means towards other vital satisfactions. If I am lost in the woods and starved, and find what looks like a cow-path, it is of the utmost importance that I should think of a human habitation at the end of it, for if I do so and follow it, I save myself. The true thought is useful here because the house which is its object is useful. The practical value of true ideas is thus primarily derived from the practical importance of their objects to us. Their objects are, indeed, not important at all times. I may on another occasion have no use for the house; and then my idea of it, however verifiable, will be practically irrelevant, and had better remain latent. Yet since almost any object may some day become temporarily important, the advantage of having a general stock of extra truths, of ideas that shall be true of merely possible situations, is obvious. We store such extra truths away in our memories, and with the overflow we fill our books *of reference. Whenever such an extra truth becomes practically relevant to one of our emergencies, it passes from cold-storage to do work in the world and our belief in it grows active. You can say of it then either that "it is useful because it is true" or that "it is true because it is useful." Both these phrases mean exactly the same thing, namely, that here is an idea that gets fulfilled and can be verified. True is the name for whatever idea starts the verification-process, useful is the name for its completed function in experience. True ideas would never have been singled out as such, would never have acquired a class-name, least of all a name suggesting value, unless they had been useful from the outset in this way.*

From this simple cue pragmatism gets her general notion of truth as something essentially bound up with the way in which one moment in our experience may lead us towards other moments which it will be worthwhile to have been led to. Primarily, and on the commonsense level, the truth of a state of mind means this function of a leading that is worthwhile. When a moment in our experience, of any kind whatever, inspires us with a thought that is true, that means that sooner or later we dip by that thought's guidance into the particulars of experience again and make advantageous connection with them. This is a vague enough statement, but I beg you to retain it, for it is essential.

Our experience meanwhile is all shot through with regularities. One bit of it can warn us to get ready for another bit, can "intend" or be "significant of" that remoter object. The object's advent is the significance's verification. Truth, in these cases, meaning nothing but eventual verification, is manifestly incompatible with waywardness on our part. Woe to him whose beliefs play fast and loose with the order which realities follow in his experience; they will lead him nowhere or else make false connections.

By "realities" or "objects" here, we mean either things of common sense, sensibly present, or else commonsense relations, such as dates, places, distances, kinds, activities. Following our mental image of a house along the cow-path, we actually come to see the house; we get the image's full verification. Such simply and fully verified leadings are certainly the originals and prototypes of the truth-process. Experience offers indeed other forms of truth-process, but they are all conceivable as being primary verifications arrested, multiplied, or substituted one for another.

Take, for instance, yonder object on the wall. You and I consider it to be a "clock," although no one of us has seen the hidden works that make it one. We let our notion pass for true without attempting to verify. If truths mean verification-process essentially, ought we then to call such unverified truths as this abortive? No, for they form the overwhelmingly large number of truths we live by. Indirect as well as direct verifications pass muster. Where circumstantial evidence is sufficient, we can go without eyewitnessing. Just as we here assume Japan to exist without ever having been there, because it works to do so, everything we know conspiring with the belief, and nothing interfering, so we assume that thing to be a clock. We use it as a clock, regulating the length of our lecture by it. The verification of the assumption here means its leading to no frustration or contradiction. Verifiability of wheels and weights and pendulum is as good as verification. For one truth-process completed there are a million in our lives that function in this state of nascency. They turn us towards direct verification; lead us into the surroundings of the objects they envisage; and then, if everything runs on harmoniously, we are so sure that verification is possible that we omit it, and are usually justified by all that happens.

Truth lives, in fact, for the most part on a credit system. Our thoughts and beliefs "pass," so long as nothing challenges them, just as bank-notes pass so long as nobody refuses them. But this all points to direct face-to-face verifications somewhere, without which the fabric of truth collapses like a financial system with no cash-basis whatever. You accept my verification of one thing, I yours of another. We trade on each other's truth. But beliefs verified concretely by somebody are the posts of the whole superstructure.

Another good reason—besides economy of time—for waiving complete verification in the usual business of life is that all things exist in kinds and not singly. Our world is found once for all to have that peculiarity. So that when we have once directly verified our ideas about one specimen of a kind, we consider ourselves free to apply them to other specimens without verification. A mind that habitually discerns the kind of thing before it, and acts by the law of the kind immediately, without pausing to verify, will be a "true" mind in ninety-nine out of a hundred emergencies, proved so by its conduct fitting everything it meets, and getting no refutation.

Indirectly or only potentially verifying processes may thus be true as well as full verification-processes. They work as true processes would work, give us the same advantages, and claim our recognition for the same reasons. All this on the common-sense level of matters of fact, which we are alone considering.

But matters of fact are not our only stock in trade. Relations among purely mental ideas form another sphere where true and false beliefs obtain, and here the beliefs are absolute, or unconditional. When they are true they bear the name either of definitions or of principles. It is either a principle or a definition that 1 and 1 make 2, that 2 and 1 make 3, and so on; that white differs less from gray than it does from black; that when the cause begins to act the effect also commences. Such propositions hold of all possible "ones," of all conceivable "whites" and "grays" and "causes." The objects here are mental objects. Their relations are perceptually obvious at a glance, and no sense-verification is necessary. Moreover, once true, always true, of those same mental objects. Truth here has an "eternal" character. If you can find a concrete thing anywhere that is "one" or "white" or "gray" or an "effect," then your principles will everlastingly apply to it. It is but a case of ascertaining the kind, and then applying the law of its kind to the particular object. You are sure to get truth if you can but name the kind rightly, for your mental relations hold good of everything of that kind without exception. If you then, nevertheless, failed to get truth concretely, you would say that you had classed your real objects wrongly.

In this realm of mental relations, truth again is an affair of leading. We relate one abstract idea with another, framing in the end great systems of logical and mathematical truth, under the respective terms of which the sensible facts of experience eventually arrange themselves, so that our eternal truths hold good of realities also. This marriage of fact and theory is endlessly fertile. What we say is here already true in advance of special verification, if we have subsumed our objects rightly. Our ready-made ideal framework for all sorts of possible objects follows from the very structure of our thinking. We can no more play fast and loose with these abstract relations than we can do so with our sense-experiences. They coerce us; we must treat them consistently, whether or not we like the results. The rules of addition apply to our debts as rigorously as to our assets. The hundredth decimal of π, the ratio of the circumference to its diameter, is predetermined ideally now, though no one may have computed it. If we should ever need the figure in our dealings with an actual circle we should need to have it given rightly, calculated by the usual rules; for it is the same kind of truth that those rules elsewhere calculate.

Between the coercions of the sensible order and those of the ideal order, our mind is thus wedged tightly. Our ideas must agree with realities, be such realities concrete or abstract, be they facts or be they principles, under penalty of endless inconsistency and frustration.

So far, intellectualists can raise no protest. They can only say that we have barely touched the skin of the matter.

Realities mean, then, either concrete facts, or abstract kinds of things and relations perceived intuitively between them. They furthermore and thirdly mean, as things that new ideas of ours must no less take account of the whole body of other truths already in our possession. But what now does "agreement" with such threefold realities mean?—to use again the definition that is current.

Here it is that pragmatism and intellectualism begin to part company. Primarily, no doubt, to agree means to copy, but we saw that the mere word "clock" would do instead of a mental picture of its works, and that of many realities our ideas can only be symbols and not copies. "Past time," "power," "spontaneity"—how can our mind copy such realities?

To "agree" in the widest sense with a reality can only mean to be guided either straight up to it or into its surroundings, or to be put into such working touch with it as to handle either it or something connected with it better than if we disagreed. Better either intellectually or practically! And often agreement will only mean the negative fact that nothing contradictory from the quarter of that reality comes to interfere with the way in which our ideas guide us elsewhere. To copy a reality is, indeed, one very important way of agreeing with it, but it is far from being essential. The essential thing is the process of being guided. Any idea that helps us to deal, whether practically or intellectually, with either the reality or its belongings, that doesn't entangle our progress in frustrations, that fits, in fact, and adapts our life to the reality's whole setting, will agree sufficiently to meet the requirement. It will hold true of that reality.

Thus, names are just as "true" or "false" as definite mental pictures are. They set up similar verification-processes, and lead to fully equivalent practical results.

All human thinking gets discursified; we exchange ideas; we lend and borrow verifications, get them from one another by means of social intercourse. All truth thus gets verbally built out, stored up, and made available for everyone. Hence, we must talk consistently just as we must think consistently: for both in talk and thought we deal with kinds. Names are arbitrary, but once understood they must be kept too. We mustn't now call Abel "Cain" or Cain "Abel." If we do, we ungear ourselves from the whole book of Genesis, and from all its connections with the universe of speech and fact down to the present time. We throw ourselves out of whatever truth that entire system of speech and fact may embody.

The overwhelming majority of our true ideas admit of no direct or face-to-face verification—those of past history, for example as of Cain and Abel. The stream of time can be remounted only verbally, or verified indirectly by the present prolongations or effects of what the past harbored. Yet if they agree with these verbalities and effects, we can know that our ideas of the past are true. As true as past time itself was, so true was Julius Caesar, so true were antediluvian monsters all in their proper dates and settings. That past time itself was, is guaranteed by its coherence with everything that's present. True as the present is, the past was also.

Agreement thus turns out to be essentially an affair of leading—leading that is useful because it is into quarters that contain objects that are important. True ideas lead us into useful verbal and conceptual quarters as well as directly up to useful sensible termini. They lead to consistency, stability, and flowing human intercourse. They lead away from excentricity and isolation, from foiled and barren thinking. The untrammeled flowing of the leading-process, its general freedom from clash and contradiction, passes for its indirect verification; but all roads lead to Rome, and in the end and eventually, all true processes must lead to the face of directly verifying sensible experiences somewhere, which somebody's ideas have copied.

Such is the large loose way in which the pragmatist interprets the word "agreement." He treats it altogether practically. He lets it cover any process of conduction from a present idea to a future terminus, provided only it run prosperously. It is only thus that "scientific" ideas, flying as they do beyond common sense, can be said to agree with their realities. It is, as I have already said, as if reality were made of ether, atoms, or electrons, but we mustn't think so literally. The term "energy" doesn't even pretend to stand for anything "objective." It is only a way of measuring the surface of phenomena so as to string their changes on a simple formula.

Yet in the choice of these man-made formulas we cannot be capricious with impunity any more than we can be capricious on the commonsense practical level. We must find a theory that will work; and that means something extremely difficult; for our theory must mediate between all previous truths and certain new experiences. It must derange common sense and previous belief as little as possible, and it must lead to some sensible terminus or other that can be verified exactly. To "work" means both these things; and the squeeze is so tight that there is little loose play for any hypothesis. Our theories are wedged and controlled as nothing else is. Yet sometimes alternative theoretic formulas are equally compatible with all the truths we know, and then we choose between them for subjective reasons. We choose the kind of theory to which we are already partial; we follow "elegance" or "economy." Clerk-Maxwell somewhere says it would be "poor scientific taste" to choose the more complicated of two equally well-evidenced conceptions; and you will all agree with him. Truth in science is what gives us the maximum possible sum of satisfactions, taste included, but consistency both with previous truth and with novel fact is always the most imperious claimant.

Analysis of James's Position

According to James, truth just is what is valuable to believe. It cannot be defined in any other way. It follows that to consider the true apart from value, as seemed to be recommended by several other positions, is both misleading and dangerous. The view of the real as objective and value-free is bound to be a delusion; our values would then merely be smuggled in, unrecognized, and we would not have the ability to compare values and direct their impact on our thought.

An adequate explanation, then, will be one that is useful. It will answer the questions we have chosen as most pressing, answers to which will make the most positive impact on our lives. The feature of this analysis that may most attract our attention is the one that most excited James: pragmatism seems, at least initially, to cut through the puzzles and dilemmas of the earlier positions. For instance, are we justified in seeing scientific explanation as a search for causes? Well, is it productive to do so? Does such a view allow us better to answer our most pressing questions? If so, by all means, we are justified in searching for causes. But can causal judgments be true? Can they be known to correspond to reality? Of course; we already answered both these questions by the assertion that the causal model of explanation is productive. There is no further appeal; no criticism beyond that judgment is possible. Any position that could claim to criticize the causal model could only justify itself by appeal to the very same productiveness, or usefulness, criterion. Should we see explanation merely as generalized description? Is is useful to do so? If this restricts our ability to answer important questions, if it is not useful in all contexts, then by all means abandon such a requirement in these contexts. If, with Scriven, we become convinced that the covering law model does not pay off, does not fill in gaps in (useful) understanding, then abandon the requirement that all scientific explanation must conform to it.

How would James probably respond to the ontological questions of Chapter 6? He was antirealist in the sense that he would accept no a priori proof that universals are real and to be discovered. Experience, to which belief must make a difference, consists in sensations and our reactions to them (second paragraph of excerpt). That sounds like nominalism. Yet James's pragmatism will not allow us to elevate nominalism into a body of restrictive doctrines that would outlaw procedures and types of explanation that might pay off. His choices among these theories were more the result of his conviction about what sorts of hypotheses tend to pay off than of any argument about what reality must be like.

Nowhere is the pragmatists' tolerance more noticeable than in their treatment of the questions of materialism and idealism. James throughout his career argued with the Hegelian idealists—but not because he thought their position was wrong. He believed much of their work to be worthless in that he thought it turned out not to make much difference whether one believed them or not. What remained of significance in idealism he believed led to multiple confusions and dilemmas. It produced consequences that could not be made consistent with other highly useful beliefs. Did James then affirm another alternative, such as materialism or dualism? He could adopt either one when it proved useful. This criterion of usefulness naturally leads one toward a multiple-aspect view of reality. In this light, we should review James's borrowed analogy (page 117) in which the pragmatic criteria serve as the common corridor through which many different ontological frameworks must pass. These different points of view need not battle one another as long as they each recognize the necessity of justifying their own procedures and assumptions in terms of payoffs in experience.

Quantum physics in our time provides a good example of a system of explanation that is acceptable on pragmatist grounds but unacceptable from some other ontological perspectives. If you ask a quantum physicist whether light is a wave phenomenon or simply particles in motion, the physicist will respond with a question of his or her own—a pragmatic one: "What do you want to know? What predictions do you want to make?" When you answer this question, the physicist will select the model appropriate to the generation of the relevant information. James, one feels sure, would have applauded this procedure. For him, it would be a waste of time to interrupt our research with questions such as Is light *really* a particle or is it a wave phenomenon?

Yet other thinkers would be, and in fact have been, profoundly disturbed with such a procedure. If science has the task of discovering what reality is like, and if we see that reality as "out there," independent of what we *want* it to be (what particular questions we want answered at the moment), then the practice-related formulae of quantum physics will seem, at best, temporary way stations in the search for the *real* nature of light. Light can't be both a collection of particles *and* at the same time a nonparticulate wave. The materialist will urge that insofar as it exhibits wave properties, the waves must be waves *of something.* Idealists may insist that waves need not be material at all—that they may be considered as pure processes in a universe that, properly speaking, contains no matter. Everyone representing materialism, idealism, and dualism can agree, however, that the noncommitment of quantum theory on this question makes all of its explanations less adequate.[2] One should notice here that critics acknowledge the *predictiveness* of quantum mechanics. But they hold that predictiveness alone is inadequate to qualify the system as explanatory.

Pragmatists do not need to reject the reality criteria we discussed in Chapter 6. However, they can suggest that to observability, simplicity, permanence, causality, and intelligibility, we add usefulness, as a vitally important criterion. Several other of the criteria, the pragmatist argues, are themselves important just because they serve usefulness. For example, why insist that whatever we take to be real must be permanent? Might we consider permanence an important reality criterion just because permanence is useful? It is useful to believe in realities that will contain fewer dangerous surprises. In a similar way, usefulness might be used to justify the other criteria. They become "useful belief" criteria rather than reality criteria.

Criticism of the Pragmatist Position

We may want to criticize pragmatism for failing the reality criteria. How can one choose a useful explanation as adequate if it does not yield permanent and intelligible knowledge, for example? Yet we have seen that pragmatists will object to the criteria list as the final court of appeal. They believe the reality criteria are themselves to be justified on grounds of usefulness. Therefore, criticism of pragmatism simply on the grounds of the criteria themselves will not really meet the issue.

Is no criticism on the basis of shared assumptions possible, then? Critics of pragmatism believe that the position contains weaknesses that can be argued about. They suggest an examination of the logic of the argument for pragmatism.

Objection: The Charge of Relativism. One of the criticisms of positivism (Chapter 4) consisted of denying that there is a universal and uniform given in sense experience. It was charged that the positivist theory leads to relativism because sensations may vary from person to person and within the same person from time to time; so no basis of agreement can be counted on to provide for orderly scientific progress. A similar charge can be leveled at pragmatism. We know that an explanation that satisfies you may leave another student puzzled and questioning. Explanations that quieted your questions last year may lead you only to question further this year. Moreover, there are no grounds in pragmatism for assuming uniformity of a given into the future other than wishful thinking. Scriven (Chapter 5) thought that his theory could avoid relativism. He believed that all the contexts for explanation share the characteristic of demanding that gaps in the understanding be filled in. Similarly, pragmatism suggests that explanation in all contexts has the characteristic of being thought useful. Now, can we avoid relativism here: do we have any *given standards* for such usefulness? Can those who are satisfied with Velikovsky's explanation of ancient events, for example, argue or eventually come to agreement with those who remain dissatisfied with it? If not, we will find it difficult to (1) account for the broad procedural agreement we find among scientists and (2) justify any norms of procedure and explanatory adequacy to guide future agreement in science. On both counts, the pragmatic analysis of explanation fails, say its critics.

Pragmatists believe there is an answer to the charge of relativism that has been brought against them. Usefulness is the criterion for the truth and adequacy of explanations; but this need not mean that everyone's private preference, everyone's personal judgment of usefulness is the final arbiter of scientific explanation. What provides for common standards then? Pragmatists point to the historically developed (and developing) scientific community. The scientific community is a community of agreement about usefulness and therefore about the adequacy of some explanations compared to others.[3]

The appeal to the standards of the scientific community is used to answer the charge of destructive relativism, and it has also allowed pragmatists to reaffirm the human flexibility of the search for truth. These standards, it is held, may change over time. What satisfies the community of inquirers at one time may not do so later, and this openness and growth in shared standards of usefulness is as critical to the power of science as is the agreement among members of the community at any one time.

The pragmatist, then, uses the standards of the scientific community as a unifying and regularizing force, whereas others might use the idea of an objectively "given" reality. In either case, one purpose of the appeal is to allow exclusion of pseudoscience and to allow agreement in the choice of more adequate explanations that can form the basis for further discovery.

Critics of pragmatism will insist that there must be a truth beyond the agreements of the scientific community. They charge that there are many communities of agreement, but that fact doesn't make what is agreed upon true. After all, the Flat-Earth Society and the Albanian Communist party both may function as communities of inquiry. Are the pragmatists claiming that no appeal to a reality beyond arbitrary community agreement is possible? If so, all these communities can have their own truth, defined by their separate understandings of what is useful, and when they conflict there will be no further court of appeal. A political party may deem it useful to believe that the theory of evolution is false or that human personality is not related to inheritance. If the scientific community disagrees with some other community, are we to say that it all depends on the community to which one belongs and that therefore no one of these beliefs is better or truer than another?

One answer to this argument has been offered by Paul Feyerabend, who is not at all daunted by the specter of all sorts of groups claiming legitimacy for their positions.[4] Feyerabend's answer has been, "Let them fight it out." He believes that astronomers, astrologists, National Institutes of Health scientists, and Velikovskyites all need to be heard and all should have access to public support. None can claim to be right on a priori grounds, and only through public debate will the truth emerge. The explanations and theories that work, that is, prove useful, are the ones that will survive. Armed with this response, pragmatists may feel secure against critics who charge them with relativism. Note, however, that granting equivalent marketplace status to any group asserting its own legitimacy seems to be a long way from the "science true because it works" picture of William James.

Objection: Two Uses of Is. A second criticism of the pragmatist position suggests that a confusion of language has made that view seem more plausible than it is. The analysis is as follows. It seems reasonable to say, as James did, that the truth is useful. It is helpful, it pays off, to know the truth. We experience that relationship between truth and usefulness a thousand times a day. The problem, and potential confusion, concerns that word *is* in the statement "The truth is useful." This statement appears plausible—it "sounds right." But does it assert that (1) one feature of the truth is that it turns out to be useful, or, in other words, statements that are true also turn out to be useful? Or does the statement mean that (2) the truth is one kind of useful thing—i.e., we only know that something is true if it first shows itself to be useful?

Critics of pragmatism can argue that the plausibility of the statement "The truth is useful" derives from sense 1 above. They will insist that sense 2 is going too far and is very dangerous as well as unnecessary. The truth can be useful without having to be just a kind of usefulness. If we stick with sense 1, we can define the truth as "what corresponds to reality," for example, and still acknowledge the happy situation that it turns out to be useful too.

To this criticism, the pragmatists may respond that *they* are not confusing senses 1 and 2 above. They forthrightly affirm sense 2. Sense 1 is insufficient because there are no standards besides usefulness.

The Significance of Pragmatism

In one sense, pragmatism is a modest program, suggesting that we avoid disputes that don't make any difference and that we include reference to the community of scientific inquirers in our theory of scientific explanation. In another sense, however, pragmatism recommends a fundamentally different understanding of the meaning of truth. In this more radical sense, it stands opposed to the common ground of the earlier theories—the separation of facts and values.

The restriction of science to facts alone promises a firm direction for scientific explanation. Whether those facts are causal relations in nature or sensations given in experience, they would allow clear decisions among competing theories regardless of the interests of the scientists involved. Science itself would be a distinctive and privileged community of inquiry, set apart from special interest groups because of its single-minded pursuit of truth. Proponents of such a value-free science can be relied upon to criticize their pragmatist opponents for relativism and to worry about potential relativism when it occurs in their own theories.

Those who, like the pragmatists, want to acknowledge the role of usefulness as a guide for science do so because they believe such criteria to be inevitable. They hold that there is no purely factual "given" in experience. How much better, they urge, to recognize the vital role of values in science than to allow those values to be smuggled in, unrecognized.

In presenting the pragmatist position, we have made no concessions to moderation in order to clarify the sense in which it presents us with a radically different view of explanation. In the history of this debate, a variety of intermediate positions have been defended. The recognition, in contemporary discussions, of the relevance of values in scientific theory represents the impact pragmatism has had on the philosophy of science and the concept of explanation.

Supplementary Readings

Gardner, Martin. *The Whys of a Philosophical Scrivener.* New York: Quill, 1983.

The second essay in this interesting book is entitled "Why I Am Not a Pragmatist." Later essays are appropriate companion readings for chapters 9 through 11.

Rorty, Richard. *Consequences of Pragmatism.* Minneapolis: University of Minnesota Press, 1982.

This is not a book for the scientist who wishes to get away from "philosophical disputes [that] collapse into insignificance," for it assumes familiarity with much of the philosophy of the current century. But Rorty's purpose is to show how pragmatism is again gaining respect as a methodology suitable to both scientists and philosophers.

Rescher, Nicholas. *The Primacy of Practice*. Oxford, England: Basil Blackwell, 1973.

Here Rescher applies the concepts of pragmatism to the problems of science.

Notes

1. William James, "Pragmatism. A New Name for Some Old Ways of Thinking," lectures II and VI, New York, 1907. Reprinted from *Essays in Pragmatism* (New York: Hafner, 1948), pp. 142–168.

2. In this spirit, Albert Einstein was among those who never accepted quantum mechanics as adequate. It seemed to him to offer two different and incompatible views of the universe at the same time and therefore to be unsatisfactory as a framework for explanation.

3. For a detailed and careful development of this response, you may refer to various essays by C. S. Peirce in such collections as *The Philosophy of Peirce: Selected Writings* (London: Routledge and Kegan Paul, 1940).

4. Paul Feyerabend, "Against Method: Outline of an Anarchistic Theory of Knowledge," in *Minnesota Studies in the Philosophy of Science*, volume 4, ed. M. Radner and S. Winokur (Minneapolis: University of Minnesota Press, 1970), pp. 17ff.

/8/ MODERN THEORIES OF EXPLANATION

The Need for a Theory of Explanation

Our survey of several classical points of view on scientific explanation can be represented as giving a series of choices: causalism versus scepticism, positivism versus a denial of the givenness of perception, the covering law model versus contextualism, and ontology versus pragmatism. We have seen strengths in several of these points of view, yet each also has limitations. Is it possible to make choices that will retain the strengths of each position so as to reap a useful set of conclusions and procedures from the dissension and conflict? Perhaps so, but in attempting to do so we are faced with a problem similar to that of the scientist who tries to bring a group of apparently conflicting observational results into meaningful order. That scientist cannot, and we cannot either, simply affirm the elements we like and clump them together without making them intelligible as a connected whole. Neither can we simply ignore those arguments whose conclusions we find unattractive. It is, of course, possible, to carry bits and pieces of various points of view into one's work as a scientist. After reading Hume, we may be more careful about making causal judgments. After reading Copi, we should be far less ready to accept data as free from all theoretical presuppositions. Perhaps, like Scriven, we will come to ask that an explanation fit the context of inquiry of the questioner. But taking to heart bits and pieces of individual points of view is like getting highway directions from several different people. They may, even taken collectively, not provide enough information to get us where we're going; and they may even contradict one another. In finding our way on the highway we need a map of the entire route; for similar reasons, scientists need a unified guide to deal with explanation.

In short, what we need is a theory! It should be a guide that addresses the classical questions we have considered. It should provide us with a consistent way to compromise and consolidate earlier answers in the light of recent thinking on the issues. It should help us with practical advice in our work as scientists.

At this point, an understandable sense of frustration often begins to afflict students of the philosophy of science. How does one develop a philosophical theory of one's own? What would such a theory look like? How would one know when such a theory was "complete"? Even more to the point, how would one begin? There are no simple and obviously complete answers to the range of questions we've been considering. Yet, by forming a picture of what a candidate theory ought to look like, we can gain an enormous advantage. It will be easier to see when elements of a proposed theory contradict one another, and we will have a checklist to use in seeing if all the main questions have been answered. The objectives of this chapter are to provide a set of questions that can serve as a guide to building a theory of scientific explanation and then to try out those questions and practice devising such a theory.

Preliminary Questions: What Do We Want a Theory of Explanation to Tell Us?

In preparing to construct a theory of explanation, we first need to know what the theory should do for us—what questions it should answer. Of course, just answering questions won't be enough because it's always possible to answer them incorrectly or inadequately. But it will help us to recognize possible *candidates* for the position of most acceptable theory if we have this minimum requirement of completeness in mind. One easy way to generate a list of questions to be answered is to review the theories of explanation considered in earlier chapters. What questions were they trying to answer? Another way to gather items for our list is to think about our actual practice of science and where decisions of principle have to be made in the course of it. From these sources— previous theories and our need to know—the following list emerges.

1. First, from a theory of explanation we should expect guidance on the *role of observations and experiment*. Are observations the things we want explained? Can observations themselves explain? The theory should tell us whether explanations must predict all or most future observations. It should commit itself on whether observations are the common source from which all truth in science comes. Much of our day-to-day work in science involves exploring the relationship between scientific theories and experiments. We need a theory of explanation that will clearly define the part experiment plays in developing, applying, and testing hypotheses and will tell us how experiments are related to scientific laws.

2. The counterpart question to the one on observations concerns *the role of theories* in science. How do theories explain? Do theories carry their own criteria for explanatory adequacy, or can we criticize a whole theory as nonexplanatory?

Do theories explain even before they are confirmed? Do we always need a theory in order to explain anything?

With answers provided to these first two sets of questions, our candidate would tell us how much of what we are accepting as explanation is theory dependent and how much is data dependent.

3.   One aspect of the relation between theories and observations is important enough to be singled out for special attention. It concerns how and when an explanation ceases to be a provisional hypothesis and becomes instead *confirmed* explanation. *How and when are explanations confirmed?* An adequate answer to this question will be rich with guidance for research, and consequently it has been the subject of much scholarship. One leading philosopher of science, Karl Popper, argues that explanations are never really confirmed. Rather, they stand as hypotheses that we try our best to falsify—to find evidence against. In this role they guide scientific investigation by helping us eliminate error.[1]

An important subissue of the confirmation question involves anomalies—experimental results that do not fit the accepted pattern of laws. A theory of explanation should tell us how to deal with anomalies that seem to contradict established laws and explanations.

4.   Several modern philosophers of science have argued strongly that the way one justifies an explanation may be quite different from the way one arrives at the explanation in the first place.[2] If there really is a difference between the two, it may be quite misleading, indeed wrong and dangerous, for us to entertain possible explanations only when we already have available the justification for them. We should, then, expect a theory of explanation to suggest *how explanations are arrived at* and to show why this procedure is the same as, or different from, the activity of justifying them.

5.   Our theory of explanation should tell us *what role the scientific community ought to play in judging explanations.* We saw the influence of the scientific community in the Velikovsky affair. We also saw the importance of the concept of community in the pragmatists' defense against relativism. All students of science have at times been asked to consult their own thought and conscience, while they have at other times been in the process of initiation into a network of values and assumptions, procedures and rules that they are hardly invited to question.

6.   In Chapter 6 it was suggested that the points of view we had considered reflect certain ontological commitments or judgments about the nature of reality. Even though we are cautioned, by positivism and pragmatism, that disputes about the nature of reality are particularly apt to be confusing and indecisive, we must remain aware that theories of explanation presuppose *some* ontology or other. Some thinkers argue that ontological commitments give us drive and vision. Almost all agree, however, that unrecognized ontological commitments can blind us to alternatives. If this is true, it is important to be explicit about our theory's *ontological premises.* Our "vision" of what reality is tends to guide, and to limit, our judgment about what is possible and what explanations are initially plausible.

Among the theory's ontological commitments, two will be of special interest to us. We will want to know, first, what the theory of explanation tells us about reductionism. Does it, for example, suggest one language about one type of real entity, in which all explanations should be expressed?

Second, we will be interested in the ontological status that the theory gives to scientific laws. Are all scientific laws simply summaries of experimental findings, or are some of them held to be true by definition or by conventional agreement?

7. Recent research in the history of various scientific disciplines will lead us to the question of the general applicability of our theory of scientific explanation. That is, do all of the sciences explain in the same way, or might physicists explain phenomena in a way that is simply inappropriate for biologists? Should we actively seek unification of method or resist it? When differences in approach among the different sciences become evident, we will need to decide whether they are superficial or the result of the very nature of the sciences.

8. Finally, we are not prescientific questioners, starting with a clean knowledge slate. We find ourselves already raised in a culture where scientific tradition affects our education and can claim enormous predictive success. Of course, we need to know how a theory of explanation can help us do science better. But it is equally true that *it should account for science's success so far,* and it must not depart in a radically destructive way from the way science is now practiced. Our hope is to understand why the best scientific explanations are so good. A theory that told us that all modern scientific explanations are inadequate would be of little relevance.

We can use these questions as a guide in probing the scope and strength of any candidate theory we are assessing, and we can also use it to help us construct our own view. As a beginning, we will look at three sample theories which reflect something of the modern debate. The first, empiricism, is developed in some detail in the text. You are encouraged to use the list of questions to fill out the detail of the two others, which we present only as sketches.

Three Sample Theories of Explanation

Empiricism. Let us begin by developing a theory of explanation that will emphasize the central themes of observing, experimenting, and testing. We can call it empiricism to reflect its commitment to experience as the source of scientific knowledge.[3] Our theory will insist that all scientific propositions coming from somewhere other than experience are to be treated very skeptically until they can be tested in experience. For example, we will distrust any claim to knowledge from reason alone or from intuition or imagination.

The roots of our empiricism go a long way back in Western civilization and science. In the fourth century B.C., Aristotle was arguing—against what he took to be the otherworldliness of Plato—that the real world consists of individual, perceivable objects of everyday experience. He held that scientific knowledge is adequate only when it accounts for the appearance of those objects in our experience.[4]

We have seen variations on the central theme of "knowledge only from experience" repeated from Berkeley and Hume (Chapters 3 and 4) in the eighteenth century down to Mach and other positivists in the early twentieth century (Chapter 4).

Late in the twentieth century, empiricism is alive and well, though having made adjustments as a result of the critique of positivism earlier. Empiricism argues the need for a solid, testable, objective foundation on which science may base its claims; empiricists find little of the required stability in the constructions of theoretical science. This suggests to them the need to find the source of knowledge in public (openly testable) experience. It is a signal of their debt to pragmatism that empiricists may be expected to emphasize the practical advantages of sticking to experience. Modern empiricism claims to give a solid foundation for real knowledge, to justify an avoidance of dependence on abstract theoretical entities, and to provide a basis for applying scientific knowledge to practical problems by emphasis on experience.

The first demand we made on a theory of explanation was that it clarify the role of observation. This will be easy for empiricism, for it holds observation to be central to scientific explanation. Observations will, generally, be the sources for generating explanations in the first place. They, or the relationships among them, will constitute what scientists are called upon to explain, and they will play a crucial role in testing explanations and applying them to technological problems. As empiricists, why are we giving such prominence to the observable? The main reason is our conviction that all knowledge must come from experience and our realization that the senses are our means of access to that experience. But why do we believe that all knowledge really does come from experience? To answer this question, we will need to recognize and articulate the ontological assumptions (question 6) behind our theory.

Our opponents might suggest at this point that our empiricist ontological commitment is as follows: what we see (hear, smell, etc.) is all that is real; no reality stands behind and accounts for what appears to us. Now we and *some* of our empiricist friends might simply agree that this does adequately summarize what we take reality to be. Berkeley and Mach (Chapter 4) are happy with this position. Yet, as empiricists, we do have another alternative that, though its consequences for explanation are the same, may sound more reasonable to us. Specifically, we may say that we don't know what reality "in itself" is like but that we should restrict scientific inquiry to what we can *know*; and that, we will urge, is the manifestation of that reality in experience. If there is a reality that does not issue in experiences, we can't know anything about it anyway, according to our principles. And whatever order of nature is *knowable* will be knowable by us as a certain ordering of appearances. Therefore, we may make either the ontological claim that appearances are all that exist or the epistemological claim (concerning knowledge) that the appearances and their relationships are all that we can know.

We have *begun* to specify the ontology of our empiricist theory, but we can hardly leave the matter here. We surely do not want to label all experience, and all observations, as "the real." One of the most obvious problems in doing so is that we could not distinguish between observational errors (reading the meter at 3, when, looked at straight, it is registering 2) and "accurate" observations. Some observations (such as the oar that looks bent in the water) are misleading. We must confront the difference in the reliability of experiences by stating how we think errors are produced and how they can be avoided. Two broad alternatives are possible in settling on the kind of experience we believe can be trusted. Phenomenalism, which we have already discussed with reference to Berkeley and Mach (Chapter 4), provides one alternative. According to phenomenalism (so called because the phenomena—or sense appearances—are taken as the real), what we observe are sense perceptions, and we cannot be in error in these observations. It makes no sense, for instance, for me to think that you may be mistaken in having a pain or in seeing red. You *may* be in error, of course, in thinking that you have a pain in your foot (if it has been amputated) or in thinking that the color you see is the same as you would observe in sunlight. Errors, we would say, involve problems of assigning certain qualities and relations to the perceptions we have. Those errors are correctable publicly and pose no insuperable problem for science if we are careful. There was considerable interest in phenomenalism in early twentieth-century Britain, particularly, and philosophers explored the ways in which "sense data," the simple perceptions taken to be the building blocks of knowledge, might be referred to by simple names, avoiding error in the building up of more complex claims.[5]

The great weakness of sense-data theories is the same one discussed in relation to Berkeley's phenomenalism in Chapter 4. Specifically, when we try to name a simple perception or to locate or identify it in some way, we must use language, with all the rich and varied framework assumptions it possesses.

This weakness of phenomenalism has led some empiricists to refine their ontology and deal with the problem of locating trustworthy experience by use of an alternative called physicalism. This type of empiricist ontology takes physical objects, not their qualities (as with phenomenalism), to be the real.[6] Such a position is certainly closer to our everyday assumptions and reflects our language, which is, in terms of sense experience, rich in reference to physical things. Errors, in this view, come especially from confusing objects, assigning inappropriate qualities and relations to objects, and confusing ourselves by reference to theoretical entities that are not physical things and therefore not real. The problem facing physicalism is clear enough. Is the wooden table in front of me both real and composed of real pieces? How about those pieces and *their* parts? Finally, must we not admit that however real we want to say the table is, the unobservable (because they are too small) parts of the table are also real? As empiricists, we will have to draw the line somewhere, perhaps at the level of direct observability under a microscope (even that is indirect in a way—is that a cell or a piece of dirt on the lens?) We will certainly ban from our ontology entities alleged merely to cause observable effects, where these entities are themselves unobservable in principle.

Is there some way to avoid making ontological commitments in our empiricist theory of explanation? After all, one traditional reason for embracing empiricism, one especially important to Mach, is the elimination of abstract metaphysics from science! Yet our judgments about reality do affect our principles of explanation. The only reason we can give for rigorously sticking to observables in science is either that observables are all there *is* to nature or, alternatively, that observables are all there is to *know* reliably about nature.

Are we committed to materialism or idealism by our theory? Note that although we *may* assert that all objects are material, we *need* not. They must only be observable, and we would not *have* to say that all observables (e.g., smells, colors, etc.) are "physical." We *would* have to go on to define what is to be meant by "physical" and "observable."

Having touched on the role of observations, and ontology, we turn now to address the second question from our list: the role of theories in explanation. Here we are likely to meet with determined opposition. Our nonempiricist friends may point out a rainbow in the sky and say, "Surely it is not enough, for science, merely to say that the colors exist (phenomenalism) or that the rainbow as an object exists (physicalism). Even to reach cautious empirical conclusions, such as the frequency of a sunshine-rain combination being simultaneous with a rainbow, is hardly science. We explain the rainbow only by reference to refraction and the nature of light, which forces us to refer to (unobservable) photons. Are empiricists going to outlaw explanations that refer to photons? If so, what is left of a scientific explanation now widely accepted?"

Does this objection endanger our position seriously? Not if we are willing to adopt a view of theories that, while consistent with our empiricism, will allow us to give meaning to concepts such as photon, force, gene, etc. To do so has been the special concern of empiricists who defend the *instrumentalist* view of theories. Instrumentalists argue that while we may freely use any theoretical constructs we choose, their use is to be justified and given meaning by correlating them with the observations they are meant to explain. Theories are *instruments* for prediction or explanation (these may, as with the covering law model, be different aspects of the same activity); they are not statements about nonobservational realities. Viewed instrumentally, then, theories do not force us to make ontological commitments.[7]

To return to our rainbow, it is possible to admit that we make a number of theoretical framework assumptions even to report observing "a rainbow." But this does no damage to our empiricist principles because these theoretical propositions are themselves ultimately understandable in terms of the observations to which they correspond. Why have theories at all? As instrumentalists, we may share Mach's view of their usefulness, as discussed earlier (Chapter 4): theories provide economical summaries of past experience and imaginative proposals for future testing. This view of theories as instruments of explanation, but as possessing no independent meaning of their own, has led to much research into the logical structure of theories. In chess, the pieces possess meaning only in the wider context of royalty and medieval warfare. Chess enthusiasts interested in the game itself ignore the external meaning of the pieces and devote themselves to a study of the rules and moves permitted by the rules (rather than kingship or what it means to be a pawn). In a similar way, empiricists interested in theories may pay particular attention to the "rules" and the logical "moves" within the theory.

We may understand theories, then, to be "uninterpreted calculi,"[8] that is, systems with axioms and rules for generating theorems from those axioms, where the axioms possess no meaning of their own (thus, they are "uninterpreted"). They may be given an interpretation, an empirical meaning, by coordinating definitions and correspondence rules that relate these terms to observations. Thus, *electron* refers to nothing real, necessarily, but it is given meaning by the observations (cloud chamber tracks, ammeter readings, etc.) that are coordinated with it. Are the laws of nature (such as the law of gravitation or electromagnetism) part of the theory? As empiricists, we have several possible answers to this question. (1) We may interpret laws as generalizations from experience that possess only some inductive probability of being true. (2) We may, instead, treat them as true by "convention"—i.e., by agreement that assuming them to be true will form part of a theory that will be predictive. In this case, since we assume but do not try to prove their truth, they are to be understood merely as "instruments" in a prediction machine. (3) Finally, we might consider laws of nature as definitions (e.g., By *force* we *mean* the product of mass and acceleration). As definitions, they are, like conventions, meaningful only as part of an overall theory. They are not capable of being denied any more than any other definition can be falsified, though it remains possible that they could be abandoned if the theory they help order turns out to be unpredictive in enough crucial cases. In all instances, the crucial point of interpretation of laws of nature for empiricists is as follows: they get their truth and even their meaning by their roots in and their successful predictions of *observations*.

How, according to our empiricist theory, are explanations justified? Obviously, the crucial test is the experimental data. The most adequate explanation is the one that enjoys the most predictive—or, failing that, retrodictive[9]—success. Now this turns out to be a complicated affair because we haven't specified which observations are most important or how many bad predictions would be necessary in order to cause us to lose faith in a given explanation. For our theory to specify this fully would no doubt be impossible; there are too many peculiarities in the specific context of a given theory and its experimental inferences. We might agree, however, that this is one place where our sketch of an empiricist view of explanation would have to be amplified by further work.

We may remain empiricists and yet deny that science's central aim is to justify theories by observations. Karl Popper[10] suggests that while observations are crucial, their role is to *falsify* theories and laws. This approach avoids the difficulties in using the probability character of observations to add up to a necessary justification. It would take an infinite number of observations to confirm decisively a generalization about the future. It may take only one to refute such a generalization. In Popper's view, we don't justify. We are entitled to regard as *more* justified a law that has withstood falsification repeatedly.

In terms of our empirical principles, we can say a good deal about how explanations are arrived at, though generally this topic will interest us less than how they are justified or falsified. We might try getting those hypotheses we wish to test only from generalizations of past experience. But even such doctrinally pure empiricists as Mach have rejected this highly restrictive method, and it is all too clear that many of the most famous scientific theories were not arrived at by simple generalization. As one modern alternative, we may follow Mach in believing that hypotheses are suggested by our imaginative extension of past experience. The imagination can be relied upon to provide models and the analogies they suggest. As opposed to those who might defend a "discoverer's intuition of reality"[11] or some process of reasoning by which the structure of reality is revealed, we as empiricists will want to separate the process of discovery into two elements: (1) the legitimate generalization from past experiences and (2) the imaginative extension of past experience, which is valuable because hypotheses may prove true experimentally. No procedure other than generalization,[12] however, has any claim to validity until it is checked in experience.

The community of scientists is important for us as empiricists, for two reasons, especially. First, it represents a fund of experimental experience and generalizations that, economically, keeps us from having to generate all the experiments and make all the observations of the past. Second, it provides a forum for the public testing and mutual checking of theories and observations so that one scientist may theorize and another test, all under accepted rules of procedure that maximize honesty and objectivity. We empiricists share a common view of what the scientific community should *not* be. It should never become a group bound together by metaphysical commitments other than a reverence for facts derived from quantitative experiments. Openness to experimental results and tolerance of diverse theories should be limited only by the rules of careful observation and reporting and the willingness of defenders of all those theories to submit the theories to public testing.

The question of whether there is a single form of explanation for all sciences over all periods of time will be answered in different ways by different empiricists. The place of the theory of evolution as explanation has been hotly disputed, with some insisting that biology is essentially descriptive and without the power to predict as the other sciences do. Most biologists respond vigorously that the historical dimension of the present set of existing species and the inability to predict the direction of further speciation are not evidence that biology is a defective science but that it simply has its own forms of evidence and argumentation. Some empiricists are willing to stick to what is observable and will demand no more of biology. Others claim that the dependence on history and the inability to predict future evolutionary changes are good evidence that biology is at best an immature science—at least on the level of species character and species change.

Finally, in reviewing our position on the list of questions our theory should speak to, we must ask how close our recommendations are to the way science is now practiced and whether or not we are suggesting major reforms. We will meet many objectors on this point. One position recently argued by some historians of science[13] is that observations are, in the actual practice of science, extremely dependent on theories (we can only *observe* what our assumptions tell us to *look for*) and that empiricists have therefore been mistaken about how science proceeds normally. Some have criticized empiricism for viewing the growth of science as a steady accumulation of generalizations, always open to test and reform. They claim that it actually proceeds "normally" in puzzle-solving fashion under quite *uncriticized* assumptions for periods of time. Empiricism is, they argue, only occasionally upset by great conceptual revolutions in which a shift of assumptions is not at all crucially dependent on observations and testing.[14]

We may join many modern empiricists in responding as follows to this challenge. It is true, we may admit, that revolutions sometimes occur in science. The Darwinian, Daltonian, and Einsteinian cases are clear examples. Furthermore, we may confess that sometimes such new theories are temporarily favored for reasons other than the evidence. Yet we may insist that in the long run it is predictive success that counts and that great imaginative leaps possess no more, ever, than the *promise* of fruitfulness based on eventual testing.[15] To the extent, then, that scientific practice has resulted in acceptance of theories for nonempirical or noneconomical motives, we suggest caution and vigilance. To *endorse* such scientific practices would be a giant step backward toward superstition.

We have now developed a sketch of our empirical theory of explanation, using our list of critical questions. How can we respond to the classical choices of earlier chapters, reconciling where possible and making commitments where necessary? We may start by acknowledging our debt to nominalism. With this nominalism we have joined a commitment to observability as a primary criterion for the real (see Chapter 6). Therefore our greatest debt, as well as our greatest potential problems, lies in the closeness of our principles to those of David Hume. Along with most modern empiricists we will have to admit that causes seem excluded from our observational reach. We may continue to use the concept as a conventional one (agreed upon as a useful but unsupported rule of thought), but we can never seek the "true causes of things" or "the causes behind, and accounting for or explaining" observations. To identify with Hume on this score is not a weakness, unless we harbor a continued hope for answers to such (meaningless) questions. However, we must acknowledge our continuing inability in modern times to escape the skeptical conclusions of Hume's broader attack on induction. How can we generalize about future observations resembling past ones (so, for example, that $F = ma$ will continue to hold and be reliable) except by assuming that there is a regularity from past to future? And how can we give evidence for that regularity inductively, for that would involve saying that past regularity implies regularity in the future, which was the very connection we were trying to account for in the first place? The skeptical conclusion of this argument is that if no connections between observations are known from experience in the first place, there is no way in the world to establish them or even to render them probable. This dilemma of empiricism has led some to abandon the whole approach as leading to a denial of the possibility of any scientific knowledge at all. If we remain loyal to empiricism, we will at the very least need to acknowledge the continued existence of this problem.

As empiricists, we need not embrace Mach's positivist position, though it remains as one possibility for us.[16] Positivism has suffered most from the recent arguments that observations are theory dependent.[17] As we have seen, it also has suffered because of the related attacks on sense-data theories, on which Mach's version seemed to depend. The most moderate course open to us is to admit the dependence of observations on theoretical assumptions, but to insist that these assumptions themselves only deserve our belief when they are given meaning by rules definitionally relating them back to observations. Given this position, we will need to abandon the hope of actually achieving a value-free and theory-free language, either of pure observations or of physical things. Our revised position can, however, see such a language as the *ideal* against which we measure explanation systems.

Our empiricist principles by themselves do not force us either to accept or to reject the covering law model of explanation. We may share the enthusiasm of its proponents for the search for the logical structure of explanations, and we will certainly applaud the thesis of the covering law theorists that the crucial difference between explanations, apart from their empirical content, rests solely in their "shape" or form.

Scriven's alternative, assuming we know only as much about it as has been presented in Chapter 5, will make us, as empiricists, both interested and suspicious. We have no reason to doubt that different contexts of empirical questions might demand different types of explanation. Working through Scriven's disagreement with Hempel will illuminate this issue more clearly for us. Yet Scriven's view of explanation as "filling in gaps in the understanding" may turn out to be either empiricist or quite at odds with empiricism, depending on what is meant by it. If coming to understand a phenomenon or experimental law means seeing how it fits into a system of related phenomena or laws so that prediction and retroduction are easier, we might adopt Scriven's definition as helpful. However, if understanding involves a process of seeing the real relations behind the appearances, the "hidden structure of reality," we will of course need to dismiss the definition as leading to unanswerable questions.

Although we can acknowledge some debt to pragmatism, we will want to urge that the distinctive service that *scientific* explanation performs is to *try* to avoid value commitments as much as possible. Explanation may be what is useful, but what is useful, we would argue, is ultimately, for science, what is predictive. And human wishing, when it gets in the way of careful prediction and open observation, clouds and impedes science.

The foregoing theory of explanation is, at best, a sketch. It is not a complete answer to all the questions we might want answered. However, it can serve as an example of the sort of theory that can address itself, with consistency, to most of the *kinds* of questions we have in mind. To see the role of such a theory as clearly as possible, let us now construct a *competing* theory, born of different principles.

Realism. For many of us, there remains the conviction that science gets at reality, reveals the structure of nature, which is manifested in the sense experiences we have, but might be hidden from someone who insisted on confining knowledge to sense experience. The structure of nature, let us argue, is capable of being thought but not always *seen* directly. Among contemporary concerns, we will identify especially with the dependence of observations on theories and on the need for the continued search for a stable and reliable truth in science (again, insisting only that such a truth is accessible, not that we have it or will in a given period find it). The distance from the observationally familiar that current scientific theories force us to travel will cause us only to reaffirm the unreliability of superficial appearances and to urge continued faith in the powers of mind. After all, we had to *learn* to see that the earth is a sphere and rotates around the sun. To the charge that an abstract and theoretical science leads us far away from practical problem solving, we can answer that experience can only be understood and manipulated as the manifestation of that hidden structure we are trying to reveal. Explanation, then, is not always of the unfamiliar by the familiar. The most familiar experiences (such as the common cold) may only be explainable by seeing them as the results of heretofore unfamiliar—indeed undreamed of—realities.

In developing a *realist* theory of explanation, our cardinal tenet will be the reality of nature, which at least in principle is independent of the more transient human wishing and valuing, and which underlies and accounts for the sense experiences we have. Experiment and testing will be important in revealing nature; but since they are only *manifestations* of reality, they themselves cannot be expected to make reality intelligible any more than shadows on the wall can themselves reveal all we may want to know about the objects casting them. Realism, thus understood, is a restatement of an ontology that threads its way through Western thought from Plato to Descartes. If we include in the tradition those who see the knowledge obtainable as essentially restricted to the realm of possible *experience*, we may include Kant as well. Many modern thinkers would affirm part or all of realism as here described. Wilfred Sellars, for example, in *Science, Perception and Reality* identifies his view as "scientific realism," but we will not attempt in what follows to be faithful to his particular point of view.[18]

Notice that realism in its ontological meaning employed here is related to but not identical with the realist position concerning universals, which was discussed in Chapter 6. The two senses of the word *realism* are the result of the two different questions to which they claim to be the answers. In medieval philosophy, the question was whether universal terms name realities. Does a term such as *bigger than*, or *animal* or *blue* refer to a real feature of the universe or is it only an abstraction from the particular object or instance that we seem to observe? *Realism* is the view that claims such terms are real features of the universe. Recall from Chapter 6 how this view was opposed by a view called nominalism, which claims that universal terms are only names for abstractions from particular experiences. In recent times philosophers of science have been interested in the question of whether theoretical entities (such as electrons or quarks) that are posited by theories of material structure are real even though we cannot even in principle hope to observe them. Sellars and other *scientific realists* have claimed they are real, whereas instrumentalists, in contrast, have said that terms such as *electron* name nothing real but are only tools, or instruments, of convenience in organizing and predicting experiences. Though these two senses of the term *realism* are different, they share common roots in Plato's view that particular sense experiences are not all there is to reality and that in fact access to the real world behind the appearances requires our reason and not our senses.

As with the empiricist perspective, the key element of our realist point of view is our ontological position. What is reality like? May we look forward to eventually understanding reality as the classical mechanists did, simply as particles in motion under certain mathematically simple laws? Well, this is certainly one possibility for realists. Yet perhaps even an emphasis on particles is misplaced. The notion "elementary particle" may itself be only a useful model to make science accommodate our imagination.

Perhaps reality consists of processes and events, and to call it "material" is only forcing thought to make concessions to photography. This much at least we as realists would want to affirm: there are real relationships between individually thought of elements in nature, and these relationships operate under regular laws so as to add up to a *structure* of being. By identifying and applying truths about that structure, scientific explanation proceeds.

Kant (Chapter 3) claimed that rational knowledge can extend only to the bounds of possible experience. Need we, as realists, disagree with Kant and hold, as earlier rationalists often did, that human understanding can reach to objects of thought that need bear no relation to human experience whatsoever? As we are describing it here, realism need not make that choice. We can be realists either in the older and more absolute sense or in the Kantian way.

What of the role of observation in our realist theory of explanation? Observations will obviously not form the crucial beginning point for all scientific explanation for us as they did for the empiricists. The truth may often come to us by the power of thought. However, we will expect it to reveal itself in a pattern of experience. Experience can thus be counted on to strengthen our conviction that we really have an explanation in a given area. Watson points out with some disdain that an empiricist-minded colleague was very tentative about the acceptance of the double helix as the explanation for the replication of DNA until checking it with experimental results. Watson held, on the contrary, that the orderliness, the neatness, the power and the beauty of the conceptual scheme he had developed seemed almost to be its own guarantee that he had found the truth.[19] Nevertheless, his confidence would have been much lower had experiments consistently not come out as predicted by his model. Observations can be counted on to help confirm a theory that we believe is correct. But as realists, we will understand that the facts are not the court of last appeal because they may be persistently misinterpreted and are, after all, theory bound anyway. Ultimately our decision concerning a given explanation will be the matching of one theory against some others rather than, in any clear sense, matching a theory with the objective and independently knowable facts. Facts are important, then, but in a more subtle and complicated way, and without the final decisiveness that they have for empiricists.

The role of theories is quite different for the realist than it is for others. Theories, even those that employ the notions of unobservable particles and forces, are going to be the ultimate truth revealers of the universe. Professor Sellars points out that it is not the role of theories simply to explain the order of facts that we have already learned through experience. Rather than explain why facts *obey* certain experimental laws, theories often explain why the phenomena do *not* match, do not obey experimental laws. His point is that we are not faced with intelligible, understandable, orderly phenomena in our sense experience. Rather, we are faced with a dismaying multitude of apparently conflicting observations that can only be made orderly by seeing them as manifestations of an underlying reality that itself contains all of them.

How shall we, as realists, say how theories are confirmed? We shall have to admit at the beginning that this is a very complicated question and that clear answers will not be possible. For one thing, we will have to deny that any final appeal can be made to experimental facts. These facts are enmeshed, if not in the theory we are considering, at least in other theories. We may use a number of characteristics of good theories to count some as more or less justified than others. Following Hempel, we will look carefully at the logical coherence of the candidate theory. However, here we must be cautious too because apparent inconsistencies may merely be the result of our not having developed the theory sufficiently or defined the terms of the theory as carefully as we could have to avoid contradictions. An important further criterion for the justification of a theory is the way in which it meshes with other theories and helps us acquire an intelligible picture of the whole field. We will therefore view with suspicion theories that taken by themselves are enormously predictive but that apparently contradict other theories in the same field. The simultaneous application of particle mechanics and wave mechanics in quantum theory will, for instance, be much more disturbing to us as realists than it would be to empiricists, who don't believe that these theories are candidates for "unlocking the structure of nature."

Fertility (the ability of a theory to generate insights beyond the range of its immediate clear application) will increase our belief in a theory's truth because of our faith that the universe is interconnected and orderly as a system. Truths about parts of it ought to lead us to analogous truths about other parts. All of these criteria then—experimental results, meshing with other theories, internal consistency—will be important to us, though none will be decisive. Theories will not be finally justified or finally falsified[20] according to realists. Theories will fade in believability as these criteria go unfulfilled or come to be held more confidently as these kinds of criteria are fulfilled to greater degrees.

Some of the most interesting work in recent philosophy of science concerns our next question. How are explanations arrived at according to our realist principles? The facts alone, even the extension of these facts by the imagination, are not likely to be the sole, or even most important, source of scientific discoveries. As realists, we will point especially to periods in the history of science when there was a breakdown in the intelligibility of the scientific vision. The multiplicity of facts that we have considered concerning justifications contribute to giving scientists an overall sense of confidence that they are on the right track or a sense that the way they have been looking at the universe does not match the way it really is. Anomalies, predictions that turn out to be false, or experimental results that when matched with the theory cannot be made consistent with it are of course important, as they are for an empiricist. For the realist, an underlying belief in the knowability of a real structure behind the welter of confusing observations provides the confidence to proceed, proposing and modifying theories to make sense out of what is observed. This persistently persuasive view of scientific discovery has been defended in recent thought by Michael Polanyi.[21] As a realist, Polanyi sees scientific discovery as the intuition of the way some aspect of reality is. He goes so far as to characterize such points of discovery as times when "nature reveals itself."

The scientific community, for realists, is characterized as far more than a brotherhood held together by motives of economy. The community is in part defined by its overall acceptance of the intelligibility of the system of nature. Realists like Polanyi insist that if most scientists were not realists, the scientific community would dissolve. Their common faith in nature, which will continue to reveal itself to careful thinkers and observers, makes the community immune from relativism and from ideology. It's easy to see that for realists the scientific community comes to be almost a moral community.[22] Inherent in our view of scientific explanation is the commitment to one unified best explanation of our experience. That there is one world rather than many is first of all a faith commitment. However, especially to the extent that we are Kantian, we may also argue that the structures of human knowing confine us to a vision of what we can understand and therefore necessitates that there will be few conflicting natures with which we have to deal.

To what extent does realism fit with the history and practice of science? As realists, we would likely argue that our point of view is essential to science as it is now practiced. We noted above that Polanyi makes this point about the necessity of the realist position. Realism as a doctrine may seem less important than realism as a useful methodological hypothesis concerning the accessibility of the truth, in its unity and in our ability to know it if we persist as a community dedicated to that goal.

From the realist point of view we are also likely to see the practices as well as the theories of science converging on a unified, coherent vision of the structure of the universe. We may not demand that all the disciplines explain in the same way at any given point in time. But we'll believe that as we bring our ideas closer to a true approximation of reality, the differences will decrease in significance.

Our position as realists obviously involves a rejection of positivism and of Hume's skepticism. If we do not reaffirm the knowability of causes in nature, at least we will insist on the intelligibility of such questions. We can endorse the covering law model if we assume that the logic of our thinking processes reflects the structure of nature. We may well see causal connections in nature as analogous to what we form in our own minds when we deduce conclusions from premises. We will take a firm stand against the thesis of pragmatism, which would urge the definition of truth in terms of workability. For realists the reason the truth works is not because we have defined it that way but because it is the truth. Since our processes of knowing may, at their best, match the structure of reality, it will be no surprise that in the long run intelligible scientific theories are predictive and useful.

Relativism. Contemporary historians of science are making significant contributions to our understanding of the past *practice* of science, but they are also affecting our view of the *nature* of science. Among them, Thomas Kuhn, to whom we have referred earlier, has provoked a particularly widespread and significant response. We will briefly develop a third possible view of the nature of scientific explanation, relativism, by using some of his main ideas.[23]

Both empiricist and realist critics of Kuhn have charged him with relativism. While Kuhn has denied that label, he has given neither observation nor theory the prominence that would satisfy the empiricist or the realist. Kuhn sees the scientific community as the central factor in what counts as good science. In this sense he owes a debt to William James, who focused on whatever the scientific community found to be useful.

Kuhn is committed by his method to describe science as he thinks it has worked and as it continues to function rather than to suggest what science ought to do and to be. In a mature science, he says, we can observe periods of "normal science" that give way to a growing sense of crisis as anomalies begin to accumulate and finally to a revolution in which an old paradigm or competing paradigms are overthrown. A new period of normal science ensues under a new paradigm. By this key notion of a paradigm, Kuhn means an overarching mindset that is not fully statable but is comprised of the important theories, experiments, instruments, and modes of operating in the discipline. A paradigm defines the appropriate questions and how they are to be answered. Acceptance of a paradigm is a conversion process, not a reasoned choice—or at least it has elements of the former and is not entirely the latter. Obviously, experimental observations are involved, as are theoretical models, but a paradigm shift is a revolution and turning to some view sufficiently new that there is always some element of incommensurability between paradigms; they cannot be laid neatly against one another and compared. The shift to a new paradigm is not an evolutionary one. It is likely to occur very quickly after a crisis has developed. There appears a whole new way of doing one's science.[24]

How would we build a theory of scientific explanation on the foregoing analysis? First, we should be ready to defend ourselves from the charge of relativism in its destructive sense, i.e., from the claim that our view means any opinion or claim to truth is as good as any other. Our defense will rest on our insistence that we are reporting how scientists actually discover, report, and come to agreement on their results; we want to ignore what some may feel scientists *ought* to be doing. Questions about the role of observations, theories, confirmation, and the relations among explanations in different sciences can all be answered by carrying out appropriately designed studies of past and present scientists and their work. How do observations and theories interact? Examine scientists' laboratories and notebooks and published works and see what they did and continue to do. Our ontological commitments will be pragmatic in the sense explained by James (Chapter 7), and we will be reminded of his image of inquiry as a series of many rooms that open onto a common hallway. Our theory is one with which many practicing scientists will feel comfortable since it attempts to avoid all the philosophical wanderings of past centuries. Scientific explanation is simply what scientists find it useful to do.

In this chapter we have seen what elements we should seek in a theory of scientific explanation. Using that list, we have examined three very different views, each loyal to some tradition in philosophy and in the interpretation of science. We have provided only sketches of these theories and left their more complete development to the reader. Further reading of some of the authors mentioned and more detailed answers to the eight questions raised in the first part of this chapter would prove helpful, as would the periodical literature in which these debates are carried on.[25]

We will now move to another set of issues that will illuminate and test our candidate theories of scientific explanation. How does explanation in science relate to, complement, or compete with explanations in other fields of human inquiry that also claim to provide explanations important to all of us?

Supplementary Readings

Churchland, Paul, and Hooker, Clifford, eds. *Images of Science: Essays on Realism and Empiricism, with a Reply from Bas C. van Fraassen.* Chicago: University of Chicago Press, 1985.

According to van Fraassen, "To be an empiricist is to withhold belief in anything that goes beyond the actual, observable phenomena. . . ." The quality of the debate between van Fraassen and his critics is high and the shared respect is even higher. This book is worth the required reading effort.

Leplin, Jarrett, ed. *Scientific Realism.* Berkeley: University of California Press, 1984.

This book includes a strong introductory chapter, five principal papers from a conference on scientific realism at the University of North Carolina, and a set of essays selected to represent the current range of thinking on scientific realism. It is an excellent source on the modern debate between the empiricists and realists.

Notes

1. Karl R. Popper, *The Logic of Scientific Discovery* (New York: Basic Books, 1959).

2. We are introducing an issue—the difference between arriving at explanations and justifying them—that has been dealt with only briefly so far. We mention it here because of its recent importance in the philosophy of

science. The modern literature contains many exciting discussions of this issue. Among them are N. R. Hanson's *Patterns of Discovery*, T. S. Kuhn's *The Structure of Scientific Revolutions*, Karl Popper's *The Logic of Scientific Discovery*, and the recent collection of articles criticizing these works, in I. Lakatos and A. Musgrave, *Criticism and the Growth of Knowledge*.

3. In this section we will not be reflecting the views of any one modern empiricist. You may find views similar to the one presented here in the work of Hans Reichenbach, Adolph Grumbaum, Richard von Mises, Bas von Fraassen, and Carl Hempel.

4. Aristotle, *Physics*, in *The Basic Works of Aristotle*, ed. Richard McKeon (New York: Random House, 1941).

5. Among the varied literature on sense data, you might first look at G. E. Moore's *Philosophical Studies* (London: Routledge and Kegan Paul, 1922). H. H. Price developed and defended the sense-data theory in his book *Perception*, (London: Methuen, 1950). Among the many attacks on sense-data theory is J. L. Austin's *Sense and Sensibilia* (Oxford, England: Clarendon Press, 1962).

6. See, for example, Rudolph Carnap, *Meaning and Necessity* (Chicago: University of Chicago Press, 1956).

7. One example of such an analysis is P. W. Bridgman, *The Nature of Some of Our Physical Concepts* (New York, Philosophical Library, 1952).

8. Carnap, *Meaning and Necessity*.

9. By retrodiction we mean the inferring of experimental results that are already known as opposed to inference of future events, which constitutes prediction.

10. *The Logic of Scientific Discovery*.

11. For example, Michael Polanyi in *Science, Faith, and Society* (London: Oxford University Press, 1946).

12. By *generalization* in this sense, we mean *complete* induction.

13. For example, Kuhn, *The Structure of Scientific Revolutions*, and Hanson, *Patterns of Discovery*.

14. This argument forms Kuhn's core thesis in *The Structure of Scientific Revolutions*.

15. Notice that here we would need to investigate cases in the history of science to test our claim.

16. For a modern defense of positivism, cf. Richard von Mises, *Positivism* (New York: Braziller, 1956).

17. Students of philosophy may want to remind us that these recent arguments concerning the dependence of observations on theories have their classical roots as far back as Plato's *Republic*.

18. Wilfred Sellars, *Science, Perception and Reality* (New York: Humanities Press, 1963).

19. James D. Watson, *The Double Helix: A Personal Account of the Discovery of the Structure of DNA* (New York: Atheneum, 1968).

20. Note that we can use Popper's scheme of falsifiability (as an alternative to a theory of confirmation) as easily as empiricists can.

21. See *The Tacit Dimension* (Garden City, N.Y.: Doubleday, 1966), as well as *Science, Faith, and Society*.

22. Cf. Jacob Bronowski, *Science and Human Values* (New York, Harper & Row, 1965) for the development of the analogy between the scientific community of truth seekers and humankind at large as a potential moral community.

23. Thomas Kuhn, *The Structure of Scientific Revolutions* (Chicago: University of Chicago Press, 1972).

24. A possible example of a paradigm shift is the change that occurred in the field of animal behavior when the reproductive behavior of males and females was examined without acceptance of the myth of the coy female. This example is of further interest in that empiricists, realists, and relativists could each derive support for their views from an examination of this episode in the history of science. See Sarah Blaffer Hrdy, "Empathy, Polyandry, and the Myth of the Coy Female," in *Feminist Approaches to Science*, ed. Ruth Bleier (New York: Pergamon Press, 1986), pp. 119–146.

25. See especially *Philosophy of Science*, the journal of the Philosophy of Science Association, Department of Philosophy, Michigan State University, East Lansing, Michigan.

/9/ EXPLANATION IN HISTORY AND SCIENCE

TO CONDUCT RESEARCH in any field, one must make fundamental commitments. One begins with assumptions as to what is worthy of study, what methods can be used, what counts as reasonable evidence, and what theoretical framework will be used. These commitments can blind an investigator to new ways of looking at a field, it is true. But they also can be liberating, for they can provide directions for research, and they help screen out masses of irrelevant considerations and bogus explanations. Commitments to a method, when one is aware of them and critical about them, can enable, not disable, the researcher.

It is clear that the assumptions scientists make have often depended on far-reaching judgments about the nature of the world and about the form and possible extent of human knowledge. In Chapter 6 we saw how explanation in science relates to several of the classical issues in metaphysics, issues that are not limited to the field of natural science. Small wonder, then, that the attitudes of scientists and toward science have been intertwined with our attitudes toward other fields that involve visions of human nature and the world: religion, politics, history, law, ethics, etc. Furthermore, changing conceptions of the nature of science throughout the centuries have often led to corresponding changes in our conceptions of ourselves. These changes have usually been hotly disputed and continue to be the focus of some of the deepest conflicts in fields other than science.

The age of classical Greece (fifth century B.C.) saw a flowering of speculation about the regularity of nature, and it was commonly believed that nature is cyclical, on the model of the repeating regularity of planetary motion. With this model in hand, Plato and Aristotle formulated analogous "laws" concerning the cyclical rise and fall of city-states. The medieval church and the science that

151

was practiced within its framework of assumptions saw nature and man as both, regulated by God's orderly will. The Enlightenment in the seventeenth and eighteenth centuries, following Newton's discoveries, sought as its primary task to find in the "moral sciences" (the study of man) laws as predictive, uniform, simple, and elegant as those that had been found for particle motion and planetary motion. In more recent times, Darwinian evolutionary theory, relativity theory, and sociobiology each provided models for understanding human life and human institutions. These attempts to make our understanding of our own social life fit a model used by the natural sciences may be legitimate, or they may all be misguided. In any case, their influence has been persistent and important and reflects the close relation between our science and our view of the world in general.

Almost every one of the suggestions that we have considered in regard to scientific explanation has had parallel influence in nonscientific fields as well. Therefore, we may well be warned that as we select among the principles of explanation for doing science, we are probably making implicit commitments that will affect our view of other fields of inquiry—their methods and their legitimacy. You may welcome the extension of these commitments as guiding an understanding of other fields, but you may also be as surprised as others have been that methods and assumptions that work very well in science seem to have awkward consequences when applied elsewhere. In this chapter and chapters 10 and 11 we will consider explanation in some fields outside natural science.

It would be interesting in itself to explore in detail how the methods and assumptions of science have affected inquiry in history, religion, and other disciplines. But our purpose is to see whether explanation in other areas of thought has relevance for a theory of scientific explanation. It may well be that the commitments we wish to make in science have consequences elsewhere that we may find unacceptable.

Some would say that the lesson to be learned from the past is that only problems result from carrying over assumptions from one field to another. Such things as the misreadings of Darwin by social evolutionists and the misapplication of church authority over science are just two examples of the troubles that can result from mixing the ideas of science with other fields. But the urge to find one set of assumptions that works equally well in the natural sciences, the social sciences, and the humanities is very strong. Certainly one reason for this is simple economy. If one method for acquiring knowledge in many different fields were discoverable, we could hope for more rapid progress in each of them. Many inquirers in the last three centuries have seemed to believe that knowledge and systematic progress in the investigation of human life could be aided by using the methods that have worked so well in science.

A second and more fundamental reason is that our logic, our human reason seems already to be a unity and does not vary from field to field. The rules of correct reasoning remain the same whether we are making change in a grocery store, gathering facts about a political candidate, or studying the causes of heart disease. It seems plausible that these rules of reason, if properly applied to all fields of human research, should be uniformly effective.

Finally, as we noted in Chapter 6, the human mind persistently yearns for and seeks a unified vision of the world. Whenever two or more fundamentally different kinds of realities have been posited—matter and force, nature and spirit, individual and community—powerful theories have emerged for uniting them.

History is an unusually good example of a discipline that has seen many battles over whether a unity of vision is possible. Many have claimed history to be natural ground for extension of the method of natural science. But the resistance to this idea has been vigorous, and most historians maintain that the study of human beings demands special rules or fundamentally different assumptions. In either case, the question remains as to how this affects our theory of scientific explanation.

What is history as a discipline? Although we will see that even this question has been the subject of substantial dispute, there are some general characteristics on which almost everyone can agree and which will suffice for our purposes here. By history we shall mean the study of the *significant human* past. Not every human event in the past is of interest to the historian (someone's cold in 1713 will be mentioned only if it had connections with events of wider importance), and natural events will be relevant only insofar as they illuminate human events. This definition, though clean and straightforward, is not free of assumptions, but it is the one we will use in looking at history and explanation.[1]

Explanation in History as the Locating of Causes

As with scientific explanation, we may well begin the study of historical explanation by focusing on causes. Historical accounts often explicitly search for the crucial precipitating events that combined to produce an event. In the standard histories of the American Civil War or the breakdown of European feudalism, evidence is assembled and arguments constructed to show the special causal significance of events or conditions that the authors believe to be crucial.

Nevertheless, we must face immediately the reservations that historians themselves voice about causal claims in history. Why the reservations? Some historians believe that causal ascriptions are often or always unjustified because historical events seem to exhibit no rational pattern to support them. Herodotus could find in the major events of the Persian War chiefly the working of a fate that, while striking down human pride, hardly worked in reliable ways.

Yet for most historians, this postulate that there are no identifiable causes for events would be crippling to their method. In this respect the historian is surely like the physicist. If one begins with the assumption that an event had no causes, that it was the working of a blind fate, what criteria can one use to select significant antecedents of the event? Even the "filling in" of events that preceded what is to be explained will demand some principles of selection.

Other historians will respond to the idea of cause in history by making the more general claim that historical writing should not be understood as explaining at all. In this interpretation, historical research is simply the study of documents and artifacts from the past, which may be used along with other disciplines to provide a simple description. (Think of the positivists in science as discussed in Chapter 4.) One reason for taking this position may be found in the argument of David Hume. Among other points, Hume argued that the observation of individual events in sequence is not ever enough to establish that they are causally related. Therefore, if history brings no assumptions or tries to bring no assumptions to its study, it will have no way to generate causal or other kinds of explanatory judgments (connective judgments) from the data it collects. This position needs to be taken seriously as a challenge to us to justify our explanatory judgments, but it has not been defended or acted upon to the extent that positivism has had influence in natural science. William Dray, in his survey, *Philosophy of History*, comments:

> Such a claim [as that history does not attempt to explain], however, is belied by the most casual glance at what historians write. They constantly claim to "throw light on" or "make clear" what they are talking about, and their exposition is richly interlarded with such explanatory expressions as "since," "therefore," and "because."[2]

One further reason for resisting a causal account of historical explanation is the recognition that so many factors enter into the context for an historical event, and prediction is always difficult if not in principle impossible. Therefore, causal accounts will always appear to claim too much for what the historian can actually produce. These concerns have led to important modifications in causal theories and some alternative accounts that are as interesting for the question of explanation in the natural sciences as in history. Since they are reactions to the causal interpretation, however, we will deal with the causal model first, keeping in mind that criticisms of it have led to alternative views, which we will consider later in this chapter.

Once one has decided to identify causal explanations in historical writing, questions arise: What kind of cause? What does a causal explanation look like? Again, as in Chapter 3, we can turn to Aristotle's list of significant causal questions for a clue. Historical accounts have exhibited material, formal, efficient, and final causes, sometimes taking these as complementary enriching causal answers and sometimes focusing on one of them as more fruitful than the others. The "material" of historical events might be said to be individual humans, or social groups or classes, and a *material cause* account may therefore concentrate on human nature—aggressiveness, or striving for freedom, or "inherent corruptness"—in accounting for events. Chronicles that find the march of history forecast in the racial character of a group or the climatic conditions of a nation are locating the explanation of events in the material underlying them.

Formal cause accounts will concentrate on the structure of the period or of the institutions and often will attempt to identify laws that govern historical change. In many cases, material and formal accounts are mixed when a historian identifies the important "matter" underlying change and then formulates laws by which changes in the matter take place. Marx, for example, isolated economically defined classes as the relevant unit and formulated laws or quasi-

laws by which the dominance of one class sets up conditions for its own destruction and the subsequent dominance of another class.

A concentration on *efficient causes* has produced accounts in which the
action of individuals and a chain of prior events are causally linked with the
event in question. In efficient cause explanations, the emphasis is on the particular action that has set an event in motion.

As in natural science, the use of *final causes* in history has been the source of
major disagreements. Those who believe that final causes are crucially important to a historical account are likely to make a sharp distinction between
reasons, on the one hand, and causes of other kinds, on the other hand. As we
will see in more detail later, one form of this emphasis is to interpret historical
events in light of the thoughts, hopes, fears, and calculations that led to them.
While the use of final causes has become increasingly out of fashion in natural
science, it continues to be an important component of historical work. The
continuing emphasis on final causes in history is part of the attempt to resist
giving human beings a purely mechanical nature.

Aristotle's classification of causal accounts can help us see, then, that there
are many ways to approach the location of causes in history and that these may
be used together or emphasized separately. The rich diversity of historical accounts of and competitive location of causes for the American Civil War, for
example, is in part due to this variety of causal approaches.

Even after we have arrived at some decision about which kind or combination of causes to look for, problems of selection remain. Which of the many
elements that form the context of the Civil War are to be called its causes? The
institution of slavery? The failure of a settled set of compromises concerning the
extension of slavery in the new lands? The economic differences between an
increasingly industrial North and a predominantly agricultural South? The Lincoln election? The interference of England and France? The firing on Fort
Sumter? Now it is true that if we have selected one kind of Aristotelian cause to
identify, the field of options will be considerably narrowed. If, for instance, we
decide to look only for structural (formal) causes, we will likely assume that the
person who happened to be president was not a crucial causal feature of the
situation: the structural causes, such as economic differences in interests, or the
type of society necessitated by the institution of slavery, would have been there
anyway regardless of the people involved. But even if this is a sufficiently open-
minded way to approach the subject (and most historians would consider it
suspect), which structural features are to be emphasized?

Notice that one always faces the same kind of further selection problem in
the cases of natural science. What accounts for the strength of a given hydro-
chloric acid solution? The electronic properties of the atoms H, O, and Cl? The
structural features of liquid water? The energy and entropy changes that occur in
the preparation of this solution? The listing of causes in natural science is also
not without its difficulties.

In history, several suggestions have been made not only about how research-
ers actually do locate causes but also how they should. We can label the sug-

gestions as the (1) importance, (2) manipulability, (3) necessary condition, and (4) prediction theories.

According to the *importance* theory, historians locate those conditions that have special importance or significance for the event in question. How is this importance determined? One school of philosophers of history, whom we may call the relativists, claim that in this selection process the moral and other values of the historians become so involved that it is impossible for reasonable disagreements to take place in historical interpretation. One person's significance may be another's triviality. Opponents of the relativists' view have insisted, though, that rough measures of importance are quite objective. Whether Hitler was in a bad mood because of a disagreeable breakfast on the morning when he ordered the invasion of Poland may have had a causal influence on when the order was issued, but surely it does not compare with the settled design on more land in Hitler's vision of the future of Germany as an important cause of the invasion, whenever it took place.

One consequence of the importance view is that historians tend to deemphasize those factors that, though necessary, were persistent and regular features of the whole period or area and highlight those that were extraordinary. This might suggest emphasis on individual actors (Thomas Paine in the American Revolution, for example), but it might also point to emerging structural considerations (such as increasing industrial competition between the colonies and England).

A second theory points to those among the many conditions for an event that are producible, avoidable, or in some other sense *manipulable*. Among others, those theorists who endorse pragmatism see special virtue in this account of causal explanation. The ten-year-old who throws a baseball through a plate glass window may in this view stand accused even while protesting that the fragility of the glass was also a necessary condition. Historians have been characterized as looking especially for manipulable causes when they focus on the errors or particular viciousness of politicians or on social conditions that with foresight could have been ameliorated.

A resolve to avoid distinctions among conditions that can be charged to the subjective situation or preferences of the historian has led some theorists to concentrate on all, and only, the *necessary conditions* for an event. While this will result in including a larger number of causative factors than the selection processes cited above, it is argued that there can be rational dispute and objective evidence for the necessity of certain conditions as opposed to others. Another important reason for concentrating on necessary conditions is that they may be distinguished from conditions that might have been sufficient to precipitate the occurrence. In this way, the historian cannot be held responsible for making predictions or retrodiction of past events. Some philosophers of history believe that it is unreasonable to expect the historian to know enough even to approximate predictive success.

Several of the preceding accounts of historical causal explanations have been rooted in their objections to a powerful and persistent theory of explanation. This theory would construe all attempts at explanation, however sketchy and deficient they might be, as aiming for location of the necessary and sufficient conditions for events. In other words, to explain is to find the factors that would contribute toward making *prediction* possible. If we know the necessary and sufficient conditions, we will know the historical laws that apply, and we will be able to predict the event. We are all familiar with this kind of historical explanation. A particularly clear example appeared in the *Christian Science Monitor* of January 8, 1980:

> There is a natural law explaining the Soviet thrust into Afghanistan. . . . Starkly stated, it is that all empires will expand, absorbing unstable areas on their borders. The expansion continues until those borders are stabilized—either by reaching a natural barrier of geography or by being brought to a halt by a countervailing zone of stability.

The most popular formulation of this prediction theory of explanation has already been met with in the covering law model discussed in Chapter 5. We mentioned at that time that one of the most persuasive appeals of the covering law model is its generalizability, and this is borne out as we see it argued by Carl Hempel and others as the appropriate model for explanation in history.

In the following selection, Hempel analyzes historical explanation. The article from which the selection was taken was originally published in the *Journal of Philosophy* in 1942. Portions dealing with the fundamental description of the covering law model that duplicate the selection in Chapter 5 have been omitted.[3]

THE FUNCTION OF GENERAL LAWS IN HISTORY

It is a rather widely held opinion that history, in contradistinction to the so-called physical sciences, is concerned with the description of particular events of the past rather than with the search for general laws which might govern those events. As a characterization of the type of problem in which some historians are mainly interested, this view probably can not be denied; as a statement of the theoretical function of general laws in scientific historical research, it is certainly unacceptable. The following considerations are an attempt to substantiate this point by showing in some detail that general laws have quite analogous functions in history and in the natural sciences, that they form an indispensable instrument of historical research, and that they even constitute the common basis of various procedures which are often considered as characteristic of the social in contradistinction to the natural sciences.

By a general law, we shall here understand a statement of universal conditional form which is capable of being confirmed or disconfirmed by suitable empirical findings. The term "law" suggests the idea that the statement in question is actually well confirmed by the relevant evidence available; as this qualification is, in many cases, irrelevant for our purpose, we shall frequently use the term "hypothesis of universal form" or briefly "universal hypothesis" instead of "general law," and state the condition of satisfactory confirmation separately, if necessary. In the content of this paper, a universal hypothesis may be assumed to assert a regularity of the following type: In every case where an event of a specified kind C occurs at a certain place and time, an event of a specified kind E will occur at a place and time which is related in a specified manner to the place and time of the occurrence of the first event. (The symbols "C" and "E" have been chosen to suggest the terms "cause" and "effect," which are often, though by no means always, applied to events related by a law of the above kind.)

The main function of general laws in the natural sciences is to connect events in patterns which are usually referred to as explanation and prediction. . . .

What is sometimes called the complete description of an individual event (such as the earthquake of San Francisco in 1906 or the assassination of Julius Caesar) would require a statement of all the properties exhibited by the spatial region or the individual object involved, or the period of time occupied by the event in question. Such a task can never be completely accomplished.

A fortiori, it is impossible to explain an individual event in the sense of accounting for all its characteristics by means of universal hypotheses, although the explanation of what happened at a specified place and time may gradually be made more and more specific and comprehensive.

But there is no difference, in this respect, between history and the natural sciences: both can give an account of their subject-matter only in terms of general concepts, and history can "grasp the unique individuality" of its objects of study no more and no less than can physics or chemistry. . . .

A set of events can be said to have caused the event to be explained only if general laws can be indicated which connect "causes" and "effect" in the manner characterized above. . . .

The use of universal empirical hypotheses as explanatory principles distinguishes genuine from pseudo-explanation, such as, say, the attempt to account for certain features of organic behavior by reference to an entelechy, for whose functioning no laws are offered, or the explanation of the achievements of a given person in terms of his "mission in history," his "predestined fate," or similar notions. Accounts of this type are based on metaphors rather than laws; they convey pictorial and emotional appeals instead of insight into factual connections; they substitute vague analogies and intuitive "plausibility" for deduction from testable statements and are therefore unacceptable as scientific explanations. . . .

The customary distinction between explanation and prediction rests mainly on a pragmatical difference between the two: While in the case of an explanation, the final event is known to have happened, and its determining conditions have to be sought, the situation is reversed in the case of a prediction: here, the initial conditions are given, and their "effect"—which in the typical case, has not yet taken place—is to be determined.

In view of the structural equality of explanation and prediction, it may be said that an explanation . . . is not complete unless it might as well have functioned as a prediction: If the final event can be derived from the initial conditions and universal hypotheses stated in the explanation, then it might as well have been predicted, before it actually happened, on the basis of a knowledge of the initial conditions and the general laws. Thus, e.g., those initial conditions and general laws which the astronomer would adduce in explanation of a certain eclipse of the sun are such that they might also have served as a sufficient basis for a forecast of the eclipse before it took place.

However, only rarely, if ever, are explanations stated so completely as to exhibit this predictive character. . . . Quite commonly, the explanation offered for the occurrence of an event is incomplete. Thus, we may hear the explanation that a barn burnt down "because" a burning cigarette was dropped in the hay, or that a certain political movement has spectacular success "because" it takes advantage of widespread racial prejudices. . . .

In some instances, the incompleteness of a given explanation may be considered as inessential. Thus, e.g., we may feel that the explanation referred to in the last example could be made complete if we so desired; for we have reasons to assume that we know the kind of determining conditions and of general laws which are relevant in this context.

Very frequently, however, we encounter "explanations" whose incompleteness can not simply be dismissed as inessential. The methodological consequences of this situation will be discussed later.

The preceding considerations apply to explanation in history as well as in any other branch of empirical science. Historical explanation, too, aims at showing that the event in question was not "a matter of chance," but was to be expected in view of certain antecedent or simultaneous conditions. The expectation referred to is not prophecy or divination, but rational scientific anticipation which rests on the assumption of general laws.

If this view is correct, it would seem strange that while most historians do suggest explanations of historical events, many of them deny the possibility of resorting to any general laws in history. It is, however, possible to account for this situation by a closer study of explanation in history, as may become clear in the course of the following analysis.

In some cases, the universal hypotheses underlying a historical explanation are rather explicitly stated. . . .

Most explanations offered in history or sociology, however, fail to include an explicit statement of the general regularities they presuppose; and there seem to be at least two reasons which account for this:

First, the universal hypotheses in question frequently relate to individual or social psychology, which somehow is supposed to be familiar to everybody through his everyday experience; thus, they are tacitly taken for granted. . . .

Second, it would often be very difficult to formulate the underlying assumptions explicitly with sufficient precision and at the same time in such a way that they are in agreement with all the relevant empirical evidence available. It is highly instructive, in examining the adequacy of a suggested explanation, to attempt a reconstruction of the universal hypotheses on which it rests. Particularly, such terms as "hence," "therefore," "consequently," "because," "naturally," "obviously," etc., are often indicative of the tacit presupposition of some general law: they are used to tie up the initial conditions with the event to be explained; but that the latter was "naturally" to be expected as "a consequence" of the stated conditions follows only if suitable general laws are presupposed. Consider, for example, the statement that the Dust Bowl farmers migrate to California "because" continual drought and sandstorms render their existence increasingly precarious, and because California seems to them to offer so much better living conditions. This explanation rests on some such universal hypothesis as that populations will tend to migrate to regions which offer better living conditions. But it would *obviously be difficult accurately to state this hypotheses in the form of a general law which is reasonably well confirmed by all the relevant evidence available. Similarly, if a particular revolution is explained by reference to the growing discontent, on the part of a large part of the population, with certain prevailing conditions, it is clear that a general regularity is assumed in this explanation, but we are hardly in a position to state just what extent and what specific form the discontent has to assume, and what the environmental conditions have to be, to bring about a revolution. Analogous remarks apply to all historical explanations in terms of class struggle, economic or geographic conditions, vested interests of certain groups, tendency to conspicuous consumption, etc.: All of them rest on the assumption of universal hypotheses which connect certain characteristics of individual or group life with others; but in many cases, the content of the hypotheses which are tacitly assumed in a given explanation can be reconstructed only quite approximately.*

It might be argued that the phenomena covered by the type of explanation just mentioned are of a statistical character, and that therefore only probability hypotheses need to be assumed in their explanation, so that the question as to the "underlying general laws" would be based on a false premise. And indeed, it seems possible and justifiable to construe certain explanations offered in history as based on the assumption of probability hypotheses rather than of general "deterministic" laws, i.e., laws in the form of universal conditions. This claim may be extended to many of the explanations offered in other fields of empirical science as well. Thus, e.g., if Tommy comes down with the measles two weeks after his brother, and if he has not been in the company of other persons having the measles, we accept the explanation that he caught the disease from his brother. Now, there is a general hypothesis underlying this explanation; but it can hardly be said to be a general law to the effect that any person who has not had the measles before will get them without fail if he stays in the company of somebody else who has the measles; that a contagion will occur can be asserted only with a high probability.

Many an explanation offered in history seems to admit of an analysis of this kind: if fully and explicitly formulated, it would state certain initial conditions, and certain probability hypotheses, such that the occurrence of the event to be explained is made highly probable by the initial conditions in view of the probability hypotheses. But no matter whether explanations in history be construed as "causal" or as "probabilistic" in character, it remains true that in general the initial conditions and especially the universal hypotheses involved are not clearly indicated, and can not unambiguously be supplemented. (In the case of probability hypotheses, for example, the probability values involved will at best be known quite roughly.)

What the explanatory analysis of historical events offer is, then, in most cases not an explanation in one of the meanings developed above; but something that might be called an explanation sketch. *Such a sketch consists of a more or less vague indication of the laws and initial conditions considered as relevant, and it needs "filling out" in order to turn into a full-fledged explanation. This filling-out requires further empirical research, for which the sketch suggests the direction. (Explanation sketches are common also outside of history; many explanations in psychoanalysis, for instance, illustrate this point.)*

Obviously, an explanation sketch does not admit of an empirical test to the same extent as does a complete explanation; and yet, there is a difference between a scientifically acceptable explanation sketch and a pseudo-explanation (or a pseudo-explanation sketch). A scientifically acceptable explanation sketch needs to be filled out by more specific statements; but it points into the direction where these statements are to be found; and concrete research may tend to confirm or to inform those indications; i.e., it may show that the kind of initial conditions suggested are actually relevant; or it may reveal that factors of a quite different nature have to be taken into account in order to arrive at a satisfactory explanation. The filling-out process required by an explanation sketch will, in general, assume the form of a gradually increasing precision of the formulations involved; but at any stage of this process, those formulations will have some empirical import: it will be possible to indicate, at least roughly, what kind of evidence would be relevant in testing them, and what findings would tend to confirm them. In the case of nonempirical explanations or explanation sketches, on the other hand—say, by reference to the historical destination of a certain race, or to a principle of historical justice—the use of empirically meaningless terms makes it impossible even roughly to indicate the type of investigation that would have a bearing upon those formulations, and that might lead to evidence either confirming or informing the suggested explanation.

In trying to appraise the soundness of a given explanation, one will first have to attempt to reconstruct as completely as possible the argument constituting the explanation or the explanation sketch. In particular, it is important to realize what the underlying explaining hypotheses are, and to judge of their scope and empirical foundation. A resuscitation of the assumptions buried under the gravestones "hence," "therefore," "because," and the like will often reveal that the explanation offered is poorly founded or downright unacceptable. In many cases, this procedure will bring to light the fallacy of claiming that a large number of details of an event have been explained when, even on a very liberal interpretation, only some broad characteristics of it have been accounted for. Thus, for example, the geographic or economic conditions under which a group lives may account for certain general features of, say, its art or its moral codes; but to grant this does not mean that the artistic achievements of the group or its system of morals has thus been explained in detail; for this would imply that from a description of the prevalent geographic or economic conditions alone, a detailed account of certain aspects of the cultural life of the group can be deduced by means of specifiable general laws.

A related error consists in singling out one of several important groups of factors which would have to be stated in the initial conditions, and then claiming that the phenomenon in question is "determined" by and thus can be explained in terms of that one group of factors.

Occasionally, the adherents of some particular school of explanation or interpretation in history will adduce, as evidence in favor of their approach, a successful historical prediction which was made by a representative of their school. But though the predictive success of a theory is certainly relevant evidence of its soundness, it is important to make sure that the successful prediction is in fact obtainable by means of the theory in question. It happens sometimes that the prediction is actually an ingenious guess which may have been influenced by the theoretical outlook of its author, but which can not be arrived at by means of his theory alone. . . .

We have tried to show that in history no less than in any other branch of empirical inquiry, scientific explanation can be achieved only by means of suitable general hypotheses, or by theories, which are bodies of systematically related hypotheses. This thesis is clearly in contrast with the familiar view that genuine explanation in history is obtained by a method which characteristically distinguishes the social from the natural sciences, namely, the method of empathetic understanding: the historian, we are told, imagines himself in the place of the persons involved in the events which he wants to explain: he tries to realize as completely as possible the circumstances under which they acted, and the motives which influenced their actions; and by this imaginary self-identification with his heroes, he arrives at an understanding and thus at an adequate explanation of the events with which he is concerned.

This method of empathy is, no doubt, frequently applied by laymen and by experts in history. But it does not in itself constitute an explanation; it rather is essentially a heuristic device; its function is to suggest certain psychological hypotheses which might serve as explanatory principles in the case under consideration. Stated in crude terms, the idea underlying this function is the following: The historian tries to realize how he himself would act under the given conditions, and under the particular motivations of his heroes; he tentatively generalizes his findings into a general rule and uses the latter as an explanatory principle in accounting for the actions of the persons involved. Now, this procedure may sometimes prove heuristically helpful; but its use does not guarantee the soundness of the historical explanation to which it leads. The latter rather depends upon the factual correctness of the empirical generalizations which the method of understanding may have suggested.

Nor is the use of this method indispensable for historical explanation. A historian may, for example, be incapable of feeling himself into the role of a paranoiac historic personality, and yet he may well be able to explain certain of his actions; notably by reference to the principles of abnormal psychology. Thus, whether the historian is or is not in a position to identify himself with his historical hero is irrelevant for the correctness of his explanation; what counts is the soundness of the general hypotheses involved, no matter whether they were suggested by empathy or by a strictly behavioristic procedure. Much of the appeal of the "method of understanding" seems to be due to the fact that it tends to present the phenomena in question as somehow "plausible" or "natural" to us; this is often done by means of attractively worded metaphors. But the kind of "understanding" thus conveyed must clearly be separated from scientific understanding. In history as anywhere else in empirical science, the explanation of a phenomenon consists in subsuming it under empirical laws; and the criterion of its soundness is not whether it appeals to our imagination, whether it is presented in suggestive analogies, or is otherwise made to appear plausible—all this may occur in pseudo-explanations as well—but exclusively whether it rests on empirically well-confirmed assumptions concerning initial conditions and general laws.

So far, we have discussed the importance of general laws for explanation and prediction, and for so-called understanding in history. Let us now survey more briefly some other procedures of historical research which involve the assumption of universal hypotheses.

Closely related to explanation and understanding is the so-called interpretation of historical phenomena in terms of some particular approach or theory. The interpretations which are actually offered in history consist either in subsuming the phenomena in question under a scientific explanation or explanation sketch; or in an attempt to subsume them under some general idea which is not amenable to any empirical test. In the former case, interpretation clearly is explanation by means of universal hypotheses; in the latter, it amounts to a pseudo-explanation which may have emotive appeal and evoke vivid pictorial associations, but which does not further our theoretical understanding of the phenomena under consideration.

Analogous remarks apply to the procedure of ascertaining the "meaning" of given historical events; its scientific import consists in determining what other events are relevantly connected with the event in question, be it as "causes," or as "effects"; and the statement of the relevant connections assumes, again, the form of explanations or explanation sketches which involve universal hypotheses; this will be seen more clearly in the subsequent section.

In the historical explanation of some social institutions great emphasis is laid upon an analysis of the development of the institution up to the stage under consideration. Critics of this approach have objected that a mere description of this kind is not a genuine explanation. This argument may be given a slightly different aspect in terms of the preceding reflections: A description of the development of an institution is obviously not simply a statement of all the events which temporally preceded it; only those events are meant to be included which are "relevant" to the formation of that institution. And whether an event is relevant to that development is not a question of the value attitude of the historian, but an objective question depending upon what is sometimes called a causal analysis of the rise of that institution. Now, the causal analysis of an event consists in establishing an explanation for it, and since this requires reference to general hypotheses, so do assumptions about relevance, and, consequently, so does the adequate analysis of the historical development of an institution. . . .

The considerations developed in this paper are entirely neutral with respect to the problem of "specifically historical laws"; neither do they presuppose a particular way of distinguishing historical from sociological and other laws, nor do they imply or deny the assumption that empirical laws can be found which are historical in some specific sense, and which are well confirmed by empirical evidence.

But it may be worth mentioning here that those universal hypotheses to which historians explicitly or tacitly refer in offering explanations, predictions, interpretations, judgments of relevance, etc., are taken from various fields of scientific research, in so far as they are not prescientific generalizations of everyday experiences. Many of the universal hypotheses underlying historical explanation, for instance, would commonly be classified as psychological, economical, sociological, and partly perhaps as historical laws; in addition, historical research has frequently to resort to general laws established in physics, chemistry, and biology. Thus, e.g., the explanation of the defeat of an army by reference to lack of food, adverse weather conditions, disease, and the like, is based on a usually tacit assumption of such laws. The use of tree rings in dating events in history rests on the application of certain biological regularities. Various methods of testing the authenticity of documents, paintings, coins, etc., make use of physical and chemical theories.

The last two examples illustrate another point which is relevant in this context: Even if a historian should propose to restrict his research to a "pure description" of the past, without any attempt at offering explanations, statements about relevance and *determination, etc., he would continually have to make use of general laws. For the object of his studies would be the past—forever inaccessible to his direct examination. He would have to establish his knowledge by indirect methods: by the use of universal hypotheses which connect his present data with those past events. This fact has been obscured partly because some of the regularities involved are so familiar that they are not considered worth mentioning at all; and partly because of the habit of relegating the various hypotheses and theories which are used to ascertain knowledge about past events, to the "auxiliary sciences" of history. Quite probably, some of the historians who tend to minimize, if not to deny, the importance of general laws for history, are actuated by the feeling that only "genuinely historical laws" would be of interest for history. But once it is realized that the discovery of historical laws (in some specified sense of this very vague notion) would not make history methodologically autonomous and independent of the other branches of scientific research, it would seem that the problem of the existence of historical laws ought to lose some of its weight.*

The remarks made in this section are but special illustrations of two broader principles of the theory of science: first, the separation of "pure description" and "hypothetical generalization and theory-construction" in empirical science is unwarranted; in the building of scientific knowledge the two are inseparably linked. And, second, it is similarly unwarranted and futile to attempt the demarcation of sharp boundary lines between the different fields of scientific research, and an autonomous development of each of the fields. The necessity, in historical inquiry, to make extensive use of universal hypotheses of which at least the overwhelming majority come from fields of research traditionally distinguished from history is just one of the aspects of what may be called the methodological unity of empirical science.

We should highlight those features of Hempel's view from the preceding selection that have been the focus of most discussion and can help us understand opposing points of view.

1.   Prediction and explanation have the same logical form. An attempt at historical explanation, however sketchy and provisional it might be, must fill in information that will help make prediction possible. Those who believe that historical research may have little to do with the prediction of future events or the retrodiction of past ones will, on this account, not be able to support such research as explaining anything at all.

2.   Since the structure of explanation/prediction in history is the same as in the sciences, the Hempel account stresses a unity of method across disciplinary lines. Insofar as different disciplines try to explain anything, they must go about it in basically the same way. Some room is left, however, to say that history is concerned with the application of general laws to the explanation of individual events, whereas the principal activity of some other disciplines (the social and natural sciences) is to generate those laws. Therefore, history is a "borrower" of generalizations from these other disciplines, and from common sense.

3.   When we say that the criterion of an explanation in history is that it works toward specifying the logically necessary and sufficient causes for the event, we eliminate one potentially large area for subjectivity and relativism in history. Logical deduction does not depend either on a historian's moral judg-

ments or on the selection of facts the historian believes important. If the necessary and sufficient conditions have been specified, an explanation has been given, and not otherwise.

4. Some thinkers, including R. G. Collingwood, who will be discussed later, have made a sharp distinction between the causes of an event that may be considered as mechanically predictable under laws, and the reasons for a historical person's actions that may be thought of as freely chosen and therefore as not subsumable under causal laws. History has been taken as concerned with reasons as opposed to the causal analysis of the sciences. However, according to the covering law model, there is no reason to make such a distinction between reasons and causes. Any antecedent condition is important in explanation only insofar as it can be connected to an event or action by a law, and in this way would be predictable, at least in principle. Therefore, to qualify as part of an explanation, reasons for an action would have to be like all other causes in this crucial respect. We should note that the covering law model does not *exclude* reasons in the mind of a historical actor as antecedent conditions of importance. The issue is whether the lawlike character of all connections envisaged by the model does not do violence to the character of reasons insofar as they are different from other causes.

Students of scientific explanation who are sympathetic to the covering law model in science may consider it only an interesting side issue that the model has been used to interpret explanation in history. However, the significance of this proposed extension is centrally important for several reasons. First, criticisms of its use in history may alert us to possible problems in its use in science. Some critics, including Michael Scriven, contend that many of their objections apply to the model wherever it is used. Second, if the model can be extended to a discipline such as history, we may be encouraged to believe that there can be found a general method that will characterize explanation in *any* field of disciplined inquiry. This would be particularly helpful in submitting to a common judge those disputes that have traditionally pitted one discipline against another—disputes concerning the predictability of human events, the significance of nonphysical or noncausal interpretations of events, etc. Third, if the model is found to be adequate as an ideal for the sciences but inappropriate or limited elsewhere, the reasons for such powers and limitations may shed important light on the fundamental perspectives of different ways of studying the world and tell us how these fundamental assumptions differ.

The covering law model has power and appeal, but it has also generated major criticisms. We will now consider four alternative views—those of Oakeshott, Scriven, the relativists, and R. G. Collingwood—that arose out of critiques of the covering law model.

History and the Unique

Michael Oakeshott, in his book *Experience and Its Modes*, argued that history and science are quite different in how they attempt to explain because science seeks understanding by generalizing, by considering an event as like other events, while history seeks understanding by looking for the unique and unrepeatable.[4] In Oakeshott's view, we come to understand a historical event, to see why it occurred, by filling in the events that occurred prior to it and finding it arising out of the context set by them. To interpret historical research as a search for or a use of universal laws would be to misunderstand that the historian is trying to understand *this* event and its unique place in a chain of events. Of course the response could be made that in calling attention to a particular *chain* the historian is making causal and lawlike assumptions about the connections of historical events with one another. Yet the emphasis Oakeshott suggested is still recognizably different from that of the covering law model. Although Oakeshott was not advocating a positivist approach that abandons the attempt to explain at all, we can note interesting parallels between his view and that of Mach, which we considered in Chapter 4. They both concentrated on discovering *what* has occurred rather than *why* it occurred, insofar as the why question involves the assertion of general laws governing classes of events. Now the view that certain disciplines seek understanding of the individual event does not make the covering law model inappropriate for science. Nor does it demand that different metaphysical commitments are needed in the two disciplines. Yet not all criticisms of the covering law model for history allow such tolerance.

Normic Generalizations, Not Covering Laws

Michael Scriven argues that the covering law model, as applied to historical explanation, is "wrong, not only in detail but in conception."[5] While many theorists claim, and Hempel would admit, that the model is in most cases an *ideal*, not a model whose conditions are met in practice, Scriven believes it is misleading even as an ideal for most cases.

Historical explanation is not all one sort of enterprise, Scriven says. While the covering law model emphasizes explanations that subsume events under physical laws, other illuminating analogues may come from "explaining the way," "explaining how something works," "explaining what something means." Sometimes historical explanation is like the unfolding events of a play in which what finally occurs is seen as a reasonable outcome or even the only possible outcome, but in which predictability is just what is avoided. In all these cases, predictiveness is, contrary to the covering law model, not sought in the explanation. An earthquake's consequences may not be predictable, but we assert with confidence that it was their cause, after the fact. Therefore, predictiveness does not seem to be necessary for all good historical explanations. That it is not *sufficient* for good historical explanation either is shown by the fact that many regularities, both physical and historical, render prediction but not explanation possible. That sunrise follows sunrise does not provide an explanation of the sunrise tomorrow, though it is highly predictable; and no one would conclude from the regular appearance of the Supreme Court justices after the session announcement, that the announcement caused their appearance.

Scriven believes the demand for deduction from laws is far too strict and too narrow because such laws are seldom precise enough, either in history or science, to allow for rigorous deduction. For a law to be known to be true, its scope must be so restricted as to become uninteresting. For example, Cortez's third expedition to Baja, California, might be marvelously explained if a general law stating that "Adventurers seeking plunder will always make a third attempt after two failures" were true. But to make the law true, we would have to narrow its scope progressively until we achieved a formulation such as "Adventurers exactly like Cortez, faced with exactly the situation and prospects he faced, will undertake a third expedition." This "law," while true, explains nothing. In the same way, scientific laws do not commonly apply exactly but only with errors or inaccuracies small enough to be of no practical concern. We try for generalizations accurate enough for our purposes but not for "true" laws that would support rigorous deductions. Further, it is often impossible even to arrange these generalizations in deductive form. In science and history both, one often learns from a mentor who like a skilled craftsperson may not be able to cite the rules of analysis and procedure under which she or he operates.

Finally, Scriven sees the demand for explanations in terms of laws as confusing an explanation with its justification (see Chapter 5). It discourages perfectly acceptable explanations by demanding that they answer unasked questions: the carpet stain is explained by saying that we knocked over an ink bottle, without our invoking the laws of physics to justify our explanation. Sometimes justifications of various kinds (truth, relevance, appropriateness) are demanded, but they are answers to another kind of question.

Fortunately, Scriven says, we don't need to search for deductions from exact laws to explain historically. Historians have available a great store of rough generalizations when a causal "why" explanation is needed (and *that* kind of explanation isn't always needed). Scriven calls these guarded generalizations "normic" statements and sees them as neither as strong as universal laws nor as weak as mere reports of statistical regularities. "Power corrupts" can serve as an example. Such statements include useful truisms, "many natural laws, some tendency statements and probability statements, and—in other areas less relevant to explaining—rules, definitions, and certain normative statements in ethics. . . . [They] can be described as norm defining; they have a *selective immunity* to apparent counterexamples."[6]

Of course, covering law advocates will insist that something fundamental has been lost in Scriven's revision. Once we substitute generalizations that have immunity from counterexamples, we risk losing objectivity in explanations. The assumptions of an age may serve to insulate it from new ways of looking at history, with no disproof available. Insofar as a statement such as "Democracy is a stage in the deterioration of aristocracy into tyranny" is protected from objective tests, it can be used to interpret events without having to marshall objective evidence in its favor. Scriven believes that explanations using normic generalizations can be as objective as needed, but in light of examples such as the above, such a claim calls for extended defense. In any case, if the Scriven position looks attractive in history, it will also look attractive in science, for the arguments are very similar.

History as Nonobjective

There have always been arguments that history is not objective, but sometimes this can only amount to a recommendation for reform and may still be consistent with the covering law model or other "objective" theories. Another position, however, finds the covering law model objectionable precisely because of its attempts at objectivity. Proponents of this view are relativists in principle. They argue that moral and conceptual framework judgments always have to be made and that these are both often unrecognized and never eliminable. Thus no objective history would ever be possible, even though within a culture or age we might strive for *comparatively* more balanced and fair accounts.

On what basis are such relativistic claims advanced? One basis arises from the scepticism about any objective sense perception or other data from experience, which we discussed in Chapter 4. If all that we experience is *interpreted* experience—if there are no bare, objective facts against which all theories may be checked—then our knowledge is always dependent on our interpretive assumptions; and there is no basis for comparing the assumptions themselves. One historian or whole culture may see the past as a movement from barbarism to the relative sophistication and civilization of the present—history as a record of progress. Another historian or culture may be convinced that history is a record of chaotic or cyclic rises and falls of order, with no evidence of progress. Seemingly random destructive events in the recent past, such as the assassination of Archduke Ferdinand before World War I, may be taken by the "chaos" view as evidence for the precipitation of major events by chance occurrences. But the "progress" view can interpret such an event as one step in eventual European unification or merely as an eddy in the overall current of progress. Some high-level personal or cultural assumptions may be idiosyncratic and harmless. But they may also dramatically and essentially influence the kind of antecedent events we look for and the kind we accept as explanations. The "progress" historian may see the assassination as an isolated contributing condition while continuing to search for the "deeper" or "real" causes of the war, whereas a "chaos" historian may weave the web of events, including the outbreak of war, around the hub of the murder. If there is no ultimate agreed upon appeal to "the facts," then the explanations of different historians or different cultures will simply be different. To say one is better will always be a claim *within* a context of assumptions and not transcending it. The covering law model itself can be seen, according to this view, as a peculiar favorite of Greek and modern European-American civilization, but of no more than restricted validity.

The debate concerning the foregoing argument for relativism is a very old one, finding early expression in Socrates' search for the truth in defiance of the sophist's implicitly relativistic claim that "man is the measure of all things" (and therefore perhaps the measures will depend on the context and assumptions of the man involved!). The most respected general counterargument against such relativism is that it commits logical suicide, as mentioned in Chapter 8. The claim that all truth is relative is false if it doesn't include itself (for then *one* truth at least would be held to be absolutely true); but if it does include itself, it is actually saying, "I believe all truth is relative, but there is no ultimately good reason I can give you for sharing my belief unless you accept my assumptions, and I can't give you objective reasons for accepting them." Irritated objectivists are also likely to insist that the facts of history, just as the facts of natural science, are objective enough to allow rational debate on almost all assumptions. Can someone from *any* culture rationally deny that there *was* an assassination and that it provoked certain reactions that led to war?

If this relativist argument fails, we should note that the covering law model of historical explanation is not thereby *established*; but it is the case that failure of the relativist argument means some objectivist account may be true.

The relativism of conceptual frameworks constitutes an attack on the objectivity and universal rationality of all explanation, including explanation in science and in history. Other relativistic arguments claim that there is an essential difference between these two disciplines because the study of human actions, and especially the study of specific human events, involves the necessity of moral judgments while science can avoid such judgments. One basis for believing that historians must use moral assessments is that they normally evaluate the *significance* of events. Such significance is not value neutral in the case of human events, since one is suggesting the good or ill effects of an action, and this involves moral and other kinds of values. We noted at the beginning of this chapter that one way of identifying causes among the antecedent conditions of an event is to look for those that are controllable or avoidable in some way. The relativist argument is an extension of that analysis, claiming that humans in fact always do locate causes in this "subjective" way.

The objections we have considered so far to the covering law model of historical explanation do not obviously call into question the metaphysical assumptions on which it can be based. Some do not even disagree on the similarity it posits between science and history. Scriven, for example, believes his objections to the model apply to both history and science, and for largely the same reasons. Relativists can either restrict their claims to the subjectivity of moral judgments or generalize them to apply to both historical and scientific inquiry.

Historical Idealism

One great tradition in the philosophy of history has argued against all theories similar to the covering law model because of a fundamental disagreement about the nature of the world. This tradition has been called *idealism* because it stresses the reality of ideas and their influence on history. We could as well characterize it as the "freedom" or "self-consciousness" school of thought, however, because the importance of ideas in human history is their role as objects of thought of self-conscious, and therefore free, human beings.

The greatest classical expression of the idealist tradition in the philosophy of history comes to us in the works of Hegel.[7] This nineteenth-century philosopher distinguished sharply between the "external" causality that natural scientists investigate and the rationality that characterizes human history. The reason for major historical changes is to be found in their role in the unfolding of progressive stages of human freedom. Earlier stages, with freedom under law only for the few, have given way through systematic and understandable changes to the greater lawful freedom of recent stages. These changes in world history have been the result of the wills of individuals—their choices. Did Hegel then believe that individuals consciously plan the growth of freedom and organize their actions to that end? No, individuals change the world by their decisions, but the results are seldom what they intended in the long run. Julius Caesar had a significant impact on the development of civilization though he may himself have been motivated by the most egoistic and shortsighted concerns. In this claim lies one of the most challenging and debatable aspects of Hegel's theory. What he termed "Spirit," and has more recently been called the "spirit of humanity," is said to work through individuals but to have a direction—a destiny—that transcends the weak and selfish motivations of those who participate in it and carry forward its designs. We use similar concepts when we refer to the "German mind," or the "American Spirit," or the "liberal imagination," although we often intend these terms more metaphorically than would fit Hegel's notion of Absolute Spirit.

A number of modern thinkers have remained skeptical about the real, transcendent existence that Hegel attributed to Spirit, but they have endorsed the distinction between natural or "external" causation and human choosing. R. G. Collingwood, a leading modern thinker in this tradition, insisted that the historian's main task is to reconstruct the rationality of what has occurred.[8] To seek rationality does not mean, according to Collingwood, to pretend that all the actions in question were carefully thought out, dispassionately analyzed, and objectively chosen. Human events are as often the result of confused and conflicting desires, unfounded fears, blind hopes. Yet what makes them human is that they were chosen; they were not simply necessary links in a determined causal chain.

The Collingwood proposal suggests specific techniques to guide historical research, which reveals how different is the historian's craft from the scientist's. One should reenact the process of thought leading up to an historical event, striving for empathy with the actors involved. To explain historically is to reveal the point, the significance of what has occurred. Idealist proposals for historical explanation, such as Collingwood's, would make the disciplines of history and natural science fundamentally different because their respective subject matters are different. The differences extend, as we have seen, to methods as well. In this view, the unity of method across disciplines is a false hope that would reduce history to a kind of pseudoscience by distorting those human actions, human choices that are history's proper concern and making them what they are not—determined sequences in a causal chain. Hempel has responded to the idealist view by acknowledging the importance of empathy and rational reconstruction. But he claims that these don't constitute explanations; instead, they are simply techniques for ferreting out antecedent conditions—thoughts in this case—and hypothesizing generalizations about them. The historian may well have to employ imaginative devices different from the scientist in gathering material for explanations; but the form of explanation itself will surely be the same—namely, a deduction from covering laws.

The Hempel rejoinder is undoubtedly an attractive one. One of the most powerful features of the covering law model is the form it sets for reasoned argument, the syllogism. Should we not say, then, that however Collingwood's historian arrives at premises, the reconstruction of the rationality of an event will surely be the arrangement of a series of premises in which the statement of the intention to act is the conclusion, and that the *form* of an explanation will be identical to the structure of the covering law model?

The idealist will reject this argument because however similar a causal chain and a chain of reasoning may be, an essential difference remains. An actor is not compelled to adopt the reasons marshaled to support a given act, but a container of gas will of necessity take up a given volume when pressure and temperature are fixed. Even when the motives have been reconstructed, we recognize that they provide reasons but do not constitute causes in the covering law sense. To reconstruct the rationality of an action does not mean that prediction or retrodiction is possible—but this predictability is an absolutely essential feature of the covering law model.

At this point, an impatient philosopher is likely to join the dispute. "Look here," she will say, "you have moved beyond what a historical explanation should be. You have a disagreement concerning a metaphysical question, and it is the metaphysical consequences of your opponents's position that bothers each of you most."

The covering law position seems to eliminate the distinction between reasons and causes by making reasons fit into a scheme of lawful predictability, just as any causes do. But for the idealists, human choice is crucial and our reasons for action are not simply a link in a hard causal chain. We are not determined by immutable laws to do all that we do. And this position troubles the determinist deeply, for then where is our hope for knowledge? If some events are the result of free choice, does this imply that there are two worlds with two different sets of rules—human actions in one, and natural events in the other? If so, how are these worlds connected?

At this point, we are no longer asking questions about historical explanation. We have raised problems about human choice that make clear that a given theory of explanation in science has deeper commitments on certain philosophic issues. In Chapter 10 we will deal with some of these larger commitments by surveying the arguments on both sides of the determinism/free will controversy.

Supplementary Readings

Theories of History. Ed. with introductions by Patrick Gardiner. Glencoe, Ill.: Free Press, 1959.

This collection of readings includes most of the important authors of the last hundred years who have written about the historical process and theories of history. Hempel, Scriven, Russell, Kant, and Nagel are some of the authors we have cited before. There is a good section on explanation and laws.

Dray, William H. *Philosophy of History.* Englewood Cliffs, N.J.: Prentice-Hall, 1964.

This is intended more for a general audience than for professional historians. An important question asked by Dray is Do the methods of the sciences apply in history? Note that in this chapter we ask the converse: Does the method of explaining in history teach us something about scientific explanation?

Notes

1. We should be clear that the definition of what in the human past is *significant* is very much a point of conflict. For example, the roles played by blacks and women have often been underplayed and sometimes entirely ignored. Another dispute has arisen between historians who focus on public figures and historians who prefer to look at the details of the lives of thousands of common folk. On a different front, Marxist historians have challenged histories that ignore the clash of economic classes. Our passing over these conflicts is not to be taken as evidence that they are inconsequential.

2. William Dray, *Philosophy of History* (Englewood Cliffs, N.J.: Prentice-Hall, 1964), p. 5.

3. Carl G. Hempel, "The Function of General Laws in History," *Journal of Philosophy* 39 (1942): 35–48.

4. Michael Oakeshott, *Experience and Its Modes* (Cambridge, England: The University Press, 1933).

5. Michael Scriven, "Truisms as Grounds for Historical Explanations," in Patrick Gardiner, ed., *Theories of History* (Glencoe, Ill.: Free Press, 1959), p. 445.

6. Ibid., p. 464.

7. See especially, G. W. F. Hegel, *Lectures on the Philosophy of History,* trans. J. Sibree, (New York, Dover, 1956).

8. R. G. Collingwood, *The Idea of History* (New York: Oxford University Press, 1946).

/10/ HUMAN FREEDOM AND SCIENTIFIC EXPLANATION

CAN HUMAN ACTIONS BE explained by the same methods we use in science to explain the behavior of molecules, planets, and spiders? Or are there limits to scientific explanation that require the use of different principles to explain human actions? Chapter 9 on historical explanation raised questions such as these, questions that generate sharply conflicting answers.

One important reason for our fascination with science has been its promised power to make our own species understandable. David Hume, in 1739, heralded in the new era of social science (or "moral science," as it was then called) by reminding his contemporaries that Newton had transformed the natural sciences a century before and that it was high time that the study of man caught up. Psychology as a discipline thus was born, in Hume's *Treatise of Human Nature*.

The century that followed saw the development of the sciences of politics, sociology, anthropology, and economics. However, progress in the social (human) sciences has been difficult and hotly disputed. Deep methodological disputes—concerning the most basic assumptions about how to proceed—mark every social science discipline.

As with history, one important source of conflict is the division between those who believe that the methods of natural science should also be used by social scientists, on the one hand, and those who believe that human actions demand different "humanistic" ways of explaining.

Involved in all of these disputes is the question of human freedom. It is the meaning of human choice that causes disagreement among historians, and human choice and freedom are inextricably linked. Those who resist the application of natural science explanatory systems to the study of humans do so because they see human freedom threatened. In this chapter we will first examine the idea of freedom and then look at the place of determinism in science. The freedom-determinism debate has a long tradition, which we will touch on, but just long enough to show its relevance for the concept of scientific explanation. Our purpose is still the same—the analysis of scientific explanation.

The Role of Freedom in the Understanding of Humans

The term *freedom* is used in many human contexts, and the confusion is multiplied by the fact that it is used not only to describe human situations but also to commend or encourage us. The "free world" refers to a political grouping of non-Communist countries, and "freedom fighters" may be revolutionaries of any political persuasion as long as they are opposed to the ruling establishment. While such uses of the term *freedom* may be helpful in articulating our political views and in rallying support, they are not of much philosophical use because of their vagueness.

Among somewhat more precise uses of the term *freedom*, it will be helpful to distinguish the following:

1. *Physical freedom* is the mere ability to move as one chooses. No assumptions are implied about the nature or conditions of the choice itself; and physical freedom is not restricted to humans. For example, a fish is physically free to swim if its fins are in working order and it is in the water, unconstrained by a stringer or the jaws of another fish. One may be physically free to move whether one actually wants to move or not.

2. *Political freedom* usually refers to the ability of citizens to choose their actions without some type of coercion. Thus, it may refer to the choice of candidates for office, where there are several candidates, not just one. Or it may mean the equal protection of the laws or freedom from arbitrary imprisonment. In each case, we have some particular form of coercion in mind and mean by *freedom* the absence of that coercion. Usually we use the term to mean freedom *from* coercion, though the term may also refer to a freedom *to* accomplish some end without impediment. For example, we may refer to some nation as free either because its citizens are free *from* arbitrary imprisonment or because they are free *to* pursue the career of their choice or free to live or travel where they wish. When used in one of these political senses, *freedom* means the same as *liberty*.

3. *Economic freedom* usually refers to the relatively unobstructed ability of individuals and groups to amass and dispose of property. The obstruction we have in mind is often government, so economic freedom can refer to the absence of governmental interference in the operation of markets—of buying and selling at prices determined by supply and demand. However, economic freedom also can refer to the absence of obstructions to the market by monopolies, cartels, unions, and other organizations whose purpose is to impose rules on exchange that go beyond supply and demand.

4. *Social freedom* is often used to mean political or economic freedom, as when we refer (vaguely) to a "free society" without specifying in what sense it is free. We may take the phrase here, however, to refer to the ability of individuals

and groups to choose actions without regard to certain social norms and rules, whether legally binding or not. For example, in the United States many minority groups have not been free to live in whatever neighborhood they choose because of majority community norms. Teenagers are likewise granted various levels of social freedom relating to hours and automobile use. Social freedom may also mean freedom from a variety of constraints such as cultural taboos, peer pressure, offensive deodorant ads, etc.

5.   *Moral freedom* is the ability to choose a morally right or wrong course of action without coercion. We should note that moral freedom does not directly depend on any of the other kinds of freedom except physical freedom. For example, a person who has no choice of political candidates may still be said to be morally free if he can steal or abstain from stealing, perform acts of kindness or refrain from them. In fact one can even have been denied physical freedom of some kinds, for example by being in jail, and still be free to treat fellow prisoners morally or immorally. Still, *some* sort of physical freedom surely is necessary for someone to be morally free. If I am chained, bound, and gagged, it hardly seems reasonable to attribute any moral freedom to me. Of what moral action, or immoral action, am I then capable?

That aspect of human freedom of most interest to us here is moral freedom; in order to make sense of moral freedom it is necessary to explore the nature of moral obligation. To choose between morally right and morally wrong acts, we must know what our moral obligations are; i.e., we must know what we ought to do or are obliged to do. First notice that we are obliged to do only what we can do. *Ought* implies *can*, but the converse is not true; *can* does not imply *ought*. Just because I can swim across a lake does not imply that I ought to. Not even a desire to do the swimming plus the ability means that I *should* swim across the lake. Capability and desire do not add up to obligation. Something more is necessary in deciding what is right that must precede choosing to do what is right. To be morally free means we can decide whether an act is right or wrong and then whether we will or won't do it.

We need to explore another aspect of moral freedom before we get to the question of how science is related to freedom. This is the association of moral responsibility with moral freedom. To be morally free does not mean that we may choose between two courses of action without regard for the consequences. Moral freedom means precisely that it makes a difference which choice we make. Knowing what we ought to do places us in the position of being responsible for the choice of doing or not doing. The difference between people and animals centers on this moral responsibility that people have. We do not hold dogs responsible for their actions in the same ways that we do people because we believe dogs do not have the freedom of choice that people do. We train dogs to obey, and we housebreak them when they are young, but they do not have the rights and responsibilities of humans. In practice, then, we withhold rights and responsibilities where we judge that there is no moral freedom.

Our response to people incapable of rational choice shows this same linking of freedom and responsibility. Infants, the mentally incompetent, and the emotionally disturbed are not held responsible for their actions, or they are given diminished responsibility in some proportion to how much their freedom of choice is diminished. In our legal system we have incorporated this same idea. Legal responsibility, whether guilt/innocence or civil liability, depends on whether a person is legally of age and of sound mind and body, that is, capable of knowing what the choices are and free to choose among them. In a court of law, a person must be shown to "know the difference between right and wrong" and must not have been coerced in order to be held legally responsible.

Is moral freedom, then, the essential defining property of humanness? Immanuel Kant (Chapter 3) would certainly agree, for he placed freedom at the center of his definition of personhood. To be a person means to be free to make moral choices.[1] This is related to but not identical with the classical Greek identification of man as the "rational animal."[2] For Aristotle, our ability to reason sets us apart from the other animals. Thus R. G. Collingwood (Chapter 9) described human history as the rational reconstruction of the past. It is, however, more than the ability to construct syllogisms; computers can do that. To be a rational animal is to choose to reason where one has the choice to reason or to not reason. It is to agree self-consciously to abide by the rules of reasoning and to accept conclusions solely on the basis of the truth of the premises that lead to them. All other potential causes for acceptance of a conclusion—whether from coercion external to us or from internal drives, or a *wish* to believe it—are to be set aside. If two friends reason together, each must assume the mutual willingness to set aside pet conclusions, prejudices, and what each has been trained to believe. But all this assumes an ability to set aside causes for belief; it assumes a freedom to suspend belief and to judge in a way that transcends the forces and motives that incline a person one way or the other. True, we don't always accomplish this free position, nor would we want to. But the possibility of such an exercise of freedom is what has been taken as essential to rationality and personhood.

The "freedom to be reasonable," then, is the common ground between humans defined as rational animals and humans defined as moral beings. Those who view *human* history as *distinctive* require that there be real freedom of choice to be reasonable or to be unreasonable, to do what is right or to do what is wrong.

Freedom and Fatalism

What conditions are necessary for humans to have moral freedom? We said earlier that it requires us to be free of coercion. To be morally free we must have real choice to act or to refrain from acting. Now a number of philosophers believe that whatever we do, we are led to do by our heredity and the influences on us in the past. If we know enough about ourselves, they hold, we will be able to predict what decisions we will make and what actions we will pursue. This view is known as *fatalism* because it suggests that we are *fated* to do what we do.[3] Is such a view compatible with moral obligation and moral freedom? These thinkers often claim that it *is* compatible. They argue that to be morally responsible for an action, you must only have actually performed the action—you must not have been coerced either by external circumstances or by someone else. If someone forces you at gunpoint to drive a stolen car, you are not legally responsible. If no such external coercion exists, and especially if you thought about the action beforehand and planned it (premeditation), then you are responsible for it and were free in all relevant senses. There is no need, these thinkers insist, to hold that your action was free from the necessities of fate. "After all," they say, "it's the *person* we are judging to be free and responsible, not the person's brain, or will, or anything else internal to the person."[4]

The position that freedom is simply freedom from external coercion has been challenged in two different ways. Consider first the problem of alcoholism and our responses to it. Some argue that alcoholism is a sin and that alcoholics are fully responsible for their besotted state and whatever acts they commit even in a drunken stupor. For those who argue this way no evidence of clouded senses or impaired judgment relieves the alcoholic of full responsibility for everything from first drink to last collapse.

At the other extreme one hears that alcoholism is a disease having physiological origin and consequences and that the predisposition to alcoholism is genetic because it clearly runs in families. One cannot then hold an alcoholic responsible for the condition any more than one can say that an arthritic is responsible for arthritis. Little about alcoholism is simple, but the key idea is that we are usually ready to assign diminished responsibility for internal coercion just as readily as we do for external coercion. To whatever extent an alcoholic is compelled to "have just one," whether by an unfortunate friend or by physiology and genes, to that extent we generally will restrict responsibility. If a person is not free from external and internal coercion, then that person is also not responsible. According to this argument, fatalism is not consistent with our generally accepted ideas of freedom and responsibility.

The second argument that fatalism is not consistent with freedom is built on the notion that to whatever extent I am fated to do something, precisely to that extent I am not free to do or not do that something. Some feel that freedom is not real freedom if I am acting as I am fated to act, that is, if I am simply responding as a rock in a gravitational field. If I choose to act kindly or to follow the laws of logic in my reasoning, then I am acting as a free person, but not if I am acting as a gum ball machine in response to inserted pennies. The meaning of acts of kindness and rationality lies essentially in the way such acts differ from gum balls.

Morals and Science

We are now closer to understanding why science and the study of moral action appear to be working at cross-purposes or even to be operating on contradictory assumptions. It is often argued that scientists are committed to fatalism as part of their very method, or even that science, if you believe its conclusions, leads directly to fatalism. These are two very different claims—that science starts by assuming fatalism or that science shows fatalism to be true—and quite possibly one of them could be true and the other false. Before we investigate them separately, however, we need to consider the plausibility of linking science and fatalism at all. After all, many scientists, particularly natural scientists, are studying molecules or interstellar forces or spiders and are not directly concerned with making scientific claims about people and morals. How can they be accused of being committed to fatalism? If they can, the answer concerns the thesis called determinism, which is often taken to characterize science and to lead to fatalism.

Determinism is the view that every fact in a given field of interest is the necessary and lawful result of some other facts, and that these facts in turn are rendered necessary by others, and so on, throughout the field. Plane geometry can be characterized as so determined because the theorems are determined by acceptance of just the axioms and the laws of logic by which we can get from the axioms to the theorems. Now, if determinism is applied to spatiotemporal events, we can say not only that facts are determined by other facts, which is true of any deductive arrangement of facts (such as geometry), but that *events* considered in the field of interest are determined by, rendered necessary by, earlier events.

So far, we are a long way from any threat to morals and freedom. The field of interest may be quite restricted and have no large implications for nature in general. The phenomena in the field of geometric optics may be taken as determined in just this way, with few or no implications outside that field of study. The rub comes, however, when we realize that the sciences are usually not seen as so restricted. Physics is the study of matter in motion—*in general*. This field includes all matter, whether organized in living bodies or in stars. A physicist may claim (though not all do) that biological organisms are simply matter in motion, arranged in a certain way. Likewise biologists may claim (though not all do) that all organisms are determined to behave the way they do by genetic and environmental laws, and that man is a complicated organism that behaves in ways that are more complex but not different in kind from the ways a spider behaves.

If this kind of general claim about the scope of a scientific discipline is made, and if the field under study is taken to be determined, then by implication human beings are also determined. Does this imply fatalism? It seems to. If all events are lawfully determined to occur merely because certain also determined past events have occurred, then any *choosing* of one action rather than another must itself be determined—and therefore it must not have been a real choice at all.

In summary, we can say that determinism alone, as we have defined it, does not imply fatalism because determinism may apply to a narrowly restricted field and may be used to make no claims beyond that field. However, *determinism* combined with *reductionism* appears to yield a fatalistic view of the nature and destiny of humankind. The concept of reductionism has its own problems, which we will discuss later, but first we need to consider the issues raised by determinism in science.

Is Determinism True?

We could answer the question Is determinism true? in two different ways. We might say that determinism is true because all the evidence from the various sciences warrants such a statement. Wherever we look in science we see proof that the world around us is determined.

A second way to argue for determinism, would be to say that it is useful, perhaps even essential, that scientists believe in determinism simply in order to do their work.

Do either of these two ways of arguing for determinism hold for science and/ or seem confirmed by science? First, consider whether the evidence convinces us that determinism is true. Right away we need to separate the question into two parts. First, is there convincing evidence for determinism in the separate sciences; second, is there convincing evidence for that combination of determinism and reductionism that gives us fatalism and the attendant consequences for human responsibility and freedom?

In physics, biology, geology, chemistry, what is the evidence for determinism? Traditionally, the evidence of classical particle mechanics was thought to be most compelling. In Newton's laws of motion we appeared to have a description of a perfectly tight causal system by which the position and velocity of a particle could be rigorously predicted on the basis of its previous states, according to laws. For this predictiveness to count as evidence for determinism, however, a number of assumptions had to be made. The prediction was applicable directly only to point masses, not to actual extended bodies in space. The extension of prediction to the latter could not be rigorous, but this could be attributed to the complexity of the mathematics involved and the difficulty of confirming measurements in the laboratory. Further, one had to be able to isolate the particle and the forces operating on it in order to study it, and one must make assumptions in the isolation process. In practice, Newtonian mechanics provided remarkable predictiveness and often led to deep confidence in the utter predictability of physical nature.

In recent times, this confidence, insofar as it is based only on the evidence provided by physics, has been shaken for some new reasons and for some reasons that were there from the beginning of the Newtonian account. What were they?

First, Hume's criticism of causal connection is still with us. Past predictability suggests future determinism *only* if we *already* believe that the future will in some respects be lawfully like the past. We may insist in response to Hume that this is a reasonable assumption or that we will assert it no matter what his argument. Or we may say with Kant that this is the way humans will always experience the world. Yet the fact remains that the *evidence alone* is not sufficient to establish determinism or even causal connections in general, lawful or not.

Second, Newtonian prediction was never without some anomalies in practice, and further work in subatomic particles has suggested that the strictly deterministic model is not alone sufficient to produce the explanation of behavior that we want to have available. Both Heisenberg's indeterminacy principle and the wave equations of quantum mechanics operate on the view that we cannot even in principle provide a strict deterministic account of the position and velocity of the smallest building blocks, whether waves or particles, of nature. Let us be clear that these observations do not show that determinism is false. They only undermine the claim that determinism is rendered true or highly probable strictly by the evidence provided by the sciences.[5]

Other fields also do not provide convincing evidence for a strict determinism. Geologists certainly seek, and believe they have in some cases found, lawlike relations between events, for example, between earth plate movement and earthquakes. In chemistry the laws of thermodynamics function with absolute precision; there is no slop in the gears. But in both cases the laws or lawlike relations are statistical in nature, and this is a substantial problem if these laws are to be taken as proof of a determined universe. If one studies the enthalpy and entropy of samples of a gas, it is a fact that the thermodynamic laws relating these properties can be shown to hold precisely, but we know that gases are made up of many, many molecules moving in all directions and at a range of speeds. How is it that random motion can yield values of enthalpy and entropy that follow the laws of thermodynamics? From a philosopher's point of view, the random behavior of individual particles utterly destroys any faith in the argument that determinism is proven by the evidence. If atoms and molecules are "governed by chance," then their behavior is not determined. And even if bunches of particles obey laws, the claim that the universe is determined is too broad a claim. The inclusion of any element of chance, and that is what is accepted every time a statistical law is asserted, means something is not known to be determined. And if some part of the universe is not known to be determined, then it is not legitimate to say that science proves determinism to be true on the evidence.

This brings us squarely to the second argument for determinism, which is that though not proven by the evidence, determinism is absolutely essential for science to be done. One powerful argument that determinism is a necessary assumption of science is as follows. Suppose we ask why a spider spins its web, and someone answers that perhaps it "just happened," that no antecedent conditions will illuminate the event. Has that person offered a scientific hypothesis? Well, how would we go about testing the hypothesis? Suppose we investigated forty different possible antecedents of web spinning, and none of them seemed always to signal that the web spinning was about to follow. Does that fact add credence to the hypothesis that the phenomenon "just happened"? Of course not. In fact, *no* evidence can be adduced in support of such a hypothesis. Is it a hypothesis at all, then? No. Rather than a possible explanation of the phenomenon, it is the claim that there is no explanation; and since no evidence could be relevant in even rendering it probable, it's really a pseudoclaim. While it looks like a hypothesis, it's really the procedural assertion that its proponent doesn't want to look any further for explanations.

Now if the analysis is correct so far, it means that the denial of relevant antecedent events is really only the refusal to seek a scientific explanation. But, in fact, we can go further. Suppose the person agreed that there were relevant antecedent conditions but asserted that they weren't connected in any lawlike way to the web spinning. That is, suppose the person claimed that although eating flies sometimes led to web spinning, and sometimes thunder precipitated the web spinning, no general rules or laws that would connect some set of prior events to web spinning in general could be found. Sometimes one thing caused it; sometimes another, and that's all one can say. Isn't that person still just giving up on scientific explanation rather than providing the framework for one?

Those who find the above argument persuasive can contend that it shows two important facts about determinism. First, determinism is necessary as an assumption for science because it directs you to look for lawful relationships rather than to give up because you believe they aren't there. Second, it is a methodological assumption rather than a conclusion of science because neither it nor its denial can be rendered probable by the evidence. Rather, it gives shape and direction for the pursuit of evidence in the first place.

Can one argue effectively that determinism is *not* a necessary methodological assumption in the sciences? Some scientists do vigorously deny that they are determinists, but it is likely that their resistance is to the more general claims that every event inside and outside their field is determined. The argument for determinism in science is a practical one. Faced with a given phenomenon in a field, it seems central to the scientific enterprise to look for antecedents that can be lawfully connected with it.

Having discussed determinism as it applies to the practice of specific natural sciences, we can now ask whether we ought to believe that the universe as a whole, including humankind, is determined. If our previous reasoning is correct, we can immediately narrow this question. Arguments have been cited to show that one should not believe that determinism is proved by the evidence produced by the sciences. But if this is the case even for the restricted claims about a particular scientific subject matter, it is equally the case with regard to the broader claim about the universe as a whole. Therefore, our question can be Is it reasonable to claim, as a *methodological* principle, that the universe, including humans, is determined?

As you might expect, we now encounter vigorous disagreement. First, consider the view of those who answer yes. They are likely to say there should really be no distinction in method between the social sciences and the natural sciences. While we can for convenience say that the social sciences study humans, insofar as they are sciences at all, they must be subject to the arguments for methodological determinism cited above. It makes as little sense, for the sake of ordered inquiry, to say that a mounting crisis, or a revolution, or a ghetto riot had no lawful antecedents as it does to say that web spinning "just happened." To deny determinism here seems, as elsewhere, to be saying that explanation is impossible. But the demand for a common methodology across all the sciences from physics to history is a much less compelling demand than the one based on the claim that the universe is proven to be determined. "I cannot manipulate men as you manipulate molecules," says the historian. It might be fun, it might even be feasible in certain cases, but in fundamental ways it is neither possible nor desirable. The very nature of the subject matter in the human disciplines shifts the way in which studies must be conducted. Thus do some social scientists reject even methodological determinism as fitting in disciplines outside the hard sciences.

Possible Solutions to the Freedom-Determinism Dispute

With the foregoing distinctions about freedom, determinism, and reductionism in mind, let's look at some of the history of the freedom-determinism debate. It is a debate that goes back to the Greeks before Plato, and the ingenuity and fierceness of the arguments indicate something of how much is thought to be at stake. Before we do this, however, there is one final clarification to make. Chance or randomness fits neither a determinist model nor a libertarian (advocate of free will) view. To say an event "just happened" is obviously not deterministic. And it should be equally clear that moral responsibility and the freedom to choose to reason do not make sense in a world where human actions "just happen." Whatever resolution of the freedom-determinism debate is discussed, there will remain the question of what to do about purely random events, or if there even are such events.

Position 1: Determinism Is Universally True, and as a Consequence, Human Freedom Is an Illusion. The main reason anyone would want to extend the deterministic thesis to reality in general, including humans, is the belief that knowing anything at all presupposes the deterministic framework. We would draw special incentives to apply it to the study of humans from all the recent progress in the social sciences. Much has been learned about human behavior by assuming its basic similarity to animal behavior. The classical assumption that humans are unique because of self-consciousness and language has been battered by studies indicating the ability of animals to use language. We may therefore feel powerfully led to affirm that differences in nature are all matters of degree and complexity, not differences in kind. At the base of our concern will be a distaste for dualism—and a desire to affirm that nature is one, is unified under regular laws.

Those who find this position persuasive will have to argue that human life does not depend on the assumption of freedom—that freedom is a hypothesis we can do without. How do we handle the problems of ascribing moral and legal responsibility and rational choice to one another? Although we may just rest with the arguments presented earlier (freedom is understood simply as freedom from *external* compulsion) some determinists have felt it necessary to go further. In one classical deterministic philosophy, Benedict de Spinoza[6] argued that we can look on humans as either active or passive. When we study them as passive, we will emphasize the conditions that lead to their behavior; when we study them as active, we will concentrate on their reason as apprehending and identifying with the laws of nature (since he argued that the laws of logic, used by reason, mirror the laws of the universe), and therefore as producing effects.

These, Spinoza held, are two aspects of the one determined universe and provide an alternative to a "two different worlds" view.

Position 2: We Are Free, and Universal Determinism Is Therefore False. Freedom is affirmed and determinism denied by those who see no way to make sense of moral and legal obligation and of reasoning (as opposed to causation) on any other hypothesis. When this line of argument is used, we should note that freedom is defended indirectly, as a necessary assumption and not as something of which we have direct awareness. The starting point is the belief that human freedom exists, and therefore people can be held responsible for their acts and can be judged guilty or innocent in a court of law. Or the proponent of this view starts with a conviction that a person can choose to hold a belief on rational grounds or can choose to believe something because it was revealed to be true. The key to this position is that choosing to reason, or accepting moral responsibility, is central, and freedom is necessary for these things to make any sense.

Some thinkers have argued a more direct defense of freedom by saying that we are immediately, intuitively aware of possessing it. They point to such cases as your deliberating about whether to raise your right arm and then raising it. Surely, they would say, you are immediately aware of your power to raise or not raise your arm! That's a direct awareness of your own freedom. For critics of this argument, Hume's analysis (see Chapter 3) may be decisive. He argued on several grounds that while you are certainly aware of deliberating and aware of raising your arm, it does not follow nor is it true that you are aware of the effectiveness of the power of your will in such a case.

If one or more of these arguments for the existence of human freedom is persuasive, we are still left with the difficult decision of how to do without the deterministic hypothesis in the social sciences. Are we to say that medicine and all the social sciences may not make the deterministic assumption at all? Perhaps we could follow Stephen Toulmin's suggestion (Chapter 3) and say that determinism is merely a useful myth, as he believes is true of causality in science. Of course it will be asked how something can be useful if it is false.

Two other avenues of interpretation have often been used by those who denied determinism but wanted to account for the successes of the human sciences. One has been to admit that humans are determined in most cases and in a majority of human behavior. When are we free? Only when we consciously choose to transcend our determinations; and that occurs only when we have chosen to act out of a recognition of our moral duty or when we have chosen to reason from premises, regardless of our predispositions to believe or not to believe a given conclusion. This line of argument is close to Kant's position. Its strength is in agreeing to determinism when we are acting from emotion, where the plausibility of determinism seems greatest. At the same time it defends the existence of freedom just at the points where moral theory seems most to demand it. This view has had vigorously to defend itself against the charge that humans don't ever act solely from duty, or at least never know that they are doing so. A defense on this point was first made in Kant's *Foundations of the*

Metaphysics of Morals, although the position itself appears to have been proposed as early as Plato's *Crito.* Since Kant, the dispute has been a rich and
interesting part of moral theory.

Finally, one might deny that determinism is applicable to human beings at
all and still agree that we humans exhibit *regularities* in our actions that make
scientific study possible. Defenders of freedom will often insist that a free person
may—or always does—have a *character* at least partly as a result of personal
choice. Free decisions over a period of years should not lead one to expect
chaos. Patterns of choice will be discernible, and even more so when large
populations rather than individuals are considered. Thus inductive science is
possible without our being forced to view past patterns as the fated result of iron
laws that determine the future as well. If we make a radical distinction between
humans and other organisms or between living and nonliving nature on this
score, we will then be saying that the regularities that science investigates will
appear to be similar, but they are not. In the one case they are the expected
regularities of deterministic laws governing phenomena, while in the latter they
are summaries of past patterns that we cannot assume to be predictive.

Position 3: Both the Determinist and Freedom Hypotheses Are Merely Methodological and Thus Do Not Compete.

It may have seemed
unnecessary to pit freedom and determinism against each other. We have been
assuming in most of the preceding analysis that both theses are neither directly
known to be true nor provable as fact. Why not make tolerance our guide? Let
science use deterministic hypotheses where needed, and let moralists, district
attorneys, and logicians assume the freedom of human beings.

The benefits of this position are obvious. What are the costs? Two important
costs deserve special attention. First, there are practical cases where a decision
between the two visions of humans must be made. These include legal cases
where guilt or innocence must be judged—and where the life of the defendant is
sometimes at stake. More commonly, we make judgments in our own lives about
what we can change by our will and what we must accept and live with. On a
societal level, as well, we make judgments of this type. Should smokers be
excluded from Medicare payments for smoking-related diseases? Are some
smokers fated to smoke, or can we expect them to assume responsibility for
their habit?

The second cost is related to the first, but it is more general. However many
methodological principles we accept, most of us strive for and operate on the
assumption that we have in some measure achieved a consistent worldview.
High-level but potentially conflicting principles introduce inconsistency, which
will be an abiding source of discomfort to us.

Position 4: There Is No Conflict If We Distinguish Between Appearance and Reality.

If we are dissatisfied with all of the previous positions, it
may be because they say both too much and too little. The first two positions
make a radical choice between freedom and determinism and seem to ignore the
importance of whichever side "lost." The third position reverses this problem

and seems excessively tolerant and indecisive. Is there any way to stake out a position that respects the claims of both sides while still asserting the truth of one position over the other? The attempt to do so for the freedom and determinism question but also to resolve a host of other apparent opposites such as change and permanence, plurality versus unity of principles, chance versus lawfulness has a long and distinguished philosophic history. The general form of the position can be stated as follows: The world does not appear to us exactly as it is. At the simplest level, we know this when we see a straight oar in the water and notice that it looks bent or when we know that a building remains the same size even when it appears to become smaller as we walk farther from it. On a deeper level, the oar appears to remain unchanged though we know it is losing some molecules from friction with sediment in the water. The building, too, is being eroded by the wind though it seems changeless. At deeper levels yet, things are not, we know, what they seem. Now let us generalize. The world as it really is may have characteristics that are quite different than the world as it appears to us. But both the real world and the world as it appears may be understandable by us. It's just that the rules for understanding the two would be different. For example, objects or events in the real world may be causally connected, whereas they may appear only to be sequential and not connected at all. All events in the real world may follow a very few laws or perhaps only one law of movement and interaction, whereas appearances may be systematically understandable only by reference to many different kinds of regularities.

How might a position such as we have described help with the freedom-determinism controversy? It might do so in either of two contrary ways.

Real Determinism and Apparent Freedom.   Some theorists say that it is reasonable to accept the real world as determined but that individuals appear to and for all practical purposes do exercise free choice. Freedom is not mere illusion, on this account, because it is a concept that allows us to understand and deal with appearances. One example of such a position is the view that God has created this world with its laws and regularities and since he is all-knowing, all-powerful, and not bound by time, he now stands outside of time seeing a completely determined world from all pasts through all futures. Human beings, however, live in time and are free to act and are responsible for their actions. When we do science, we see the regularities that exist from past into future. When we are faced with a moral choice, we are in that moment free and responsible.[7] The chief opponents of this position say that they can see no difference between it and the position that freedom is illusion and human beings are simply controlled by fate.

Real Freedom and Apparent Determinism.   Here the placement of freedom and determinism with respect to reality and appearance is reversed. For Kant, Hegel, and some later German and French thinkers, the key point is that freedom is real and science is a matter of the world as it appears to us. In Chapter 3, we mentioned Kant's attempted solution to the problems of Hume's scepticism. Kant argued that the world as it appears to us is organized by human categories and forms of intuition. The world in its appearance to us is spatial, temporal, causally connected, and has a number of other features that also define its suitability for human knowing. On the other hand, of the world of things in themselves, Kant argued, we have and can have no scientific knowledge whatever.

According to Kant, can we say anything at all about the world of things as they really are? Not for the purposes of scientific knowledge or, as Kant put it, of "speculative reason." However, we are forced to make certain claims about this world for the purposes of morals, or "practical reason." As we saw early in this chapter, Kant argued that the meaning of our existence as people involves the fact that we possess rights and obligations with respect to one another. But for these to exist, freedom must be possible. We can find this freedom nowhere in the appearances (the observable behavior) of ourselves and others, and this will throw morality and science into irreconcilable conflict with each other until we remember that science does not and cannot make claims about the world behind the appearances. For the sake of fulfilling our moral obligations to one another, however, we not only can, we *must* make claims about a free personality transcending the appearances. My own freedom is not merely an interesting topic for speculation. It is a morally necessary assumption because I have obligations to respect others as people, and to deny freedom would be to deny these obligations. That would be not scepticism but immorality.

The arguments for and against Kant's division of appearance and reality are too lengthy to present in more detail here. The reasonableness of his rigid distinction has been a central matter of philosophical debate since 1789. The importance of the position lies in its potential strength to do justice to the claims of both science and morality, and as the problems discussed in this chapter make clear, that is no small strength. Moralists may make claims about human freedom and obligations; scientists may see the world as bound by a tight causality, but what is studied is the world of appearances. However, both Kant's defenders and opponents have tended to soften the sharp appearance-reality distinction.

The impact of Kant's thinking has been especially strong on the social sciences, as for example in history, through the ideas of Hegel. From the Kantian perspective, some philosophers have offered a critique of the social sciences insisting that social science can illuminate only the way people behave or appear; not from a study of human behavior alone can we make claims about the way humans really are. But social scientists also then are defended from "humanistic social science," which is not really science at all, according to this theory. Jean Piaget thus challenged as pseudoscience the work of Sartre, Merleau-Ponty, and others who would limit the observations and experimentation of social scientists by ideas of what humans really are.

Emergentism

We have seen that attempts to deal with the freedom-determinism question operate either by denying the reality of one or the other, or by subordinating the one to the other by a distinction between fact and methodological postulate or between reality and appearance. It has been hard to conceive of them as equally true of the world. After all, if we are free and the rest of nature is determined, then how are we related to the rest of nature? We must be a radically different sort of creation. To put it another way, how would we conceive of the uniqueness of humans and of our freedom, which seems so different from nature, when all the rest of nature seems unified, subject to common laws, etc.?

One position on this question has tried to deal directly with how a free creature could come to be in a world that was originally tightly deterministic. The position is called emergentism because it holds that new properties or characteristics of some beings have emerged through evolutionary change.[8] From inorganic nature, life comes to be, and the proper description of a living organism can no longer be translated into statements solely about its inorganic parts. A new reality has emerged. So, too, as organic life became more complex, consciousness and finally self-consciousness emerged as a characteristic of a few organisms. Again, once it comes to be, self-consciousness can no longer be understood by being reduced to its merely organic beginnings. Life has laws of its own, and self-consciousness has features of its own as well. One of the striking irreducible features of a self-conscious being is the fact that it does not merely behave according to laws but also acts in accordance with its thinking about laws and can even create laws.

The emergentist position has been criticized as overly metaphysical and not properly grounded in evidence. However, its defenders have seen the theory not as a scientific conclusion as much as a way of making sense, philosophically, of a world in which freedom and determinism seem both to have a place. The appeal of the position is obvious and very great. If it is tenable, it puts both freedom and determinism in the world we experience. There is no need to make a radical distinction between appearance and reality. Real differences among beings can come to be even in a world that has been created, or has always existed, under universal laws of nature.

Scientific Explanation and Freedom-Determinism Solutions

We can now turn to a consideration of the relationships between some of the theories of scientific explanation discussed earlier and various points of view regarding freedom and determinism. This topic will be particularly helpful in understanding theories of scientific explanation because such theories are always part of a larger view of knowledge and reality (as we saw in Chapter 6). Kant is a good example. He was interested, to be sure, in putting our understanding of scientific knowledge on firm foundations. But his larger project was to do this as part of an account of the nature and possibility of morality. Both science and ethics were integral parts of his overall project. Since we have already dealt with his views in this chapter, let us now consider a few other positions that have commanded much of our attention earlier in this work.

Scientific Explanation as Locating Causes. Aristotle's classification of different types of causal explanation (Chapter 3) will remind us that all causalists are not likely to agree on the freedom-determinism issue.

1.   A commitment to the existence of human moral freedom usually comes with a belief in the importance of final cause explanations of some human actions. Final cause explanations of actions deal with the purposes they serve, the intentions behind the choices of agents. Those who believe freedom is not an illusion are likely to view final cause as not entirely reducible to the other forms of causal explanation. The converse does not hold, however. One who believes that scientific explanation is the location of cause may find final cause arguments useful and also at the same time believe purposes and intentions are as rigidly predictable and deterministic as efficient causes.

2.   Causalists who view human actions as determined do so not on the basis of evidence that determinism has been proven, for nowhere in the sciences, from psychology to physics, has that evidence been obtained. (See earlier sections of this chapter.) There is also the reductionist postulate (which may or may not be accepted) that what works for billiard balls also works for human actions. Without the reductionist step, causalists may take any of the positions listed as solutions to the freedom-determinism debate.

The Covering Law Model. The crucial variable concerning covering law models of explanation is their *scope*. We have seen that any particular science might be characterized as a system of covering laws with specification of antecedent conditions and their consequences. If all explanation is claimed to be of this type and if we insert the further (reductionist) claim that the covering laws of one field are ultimately the consequences of more fundamental covering laws of another field, we are led to determinism.

Without reductionism, the covering law model is consistent with either freedom claims or a deterministic view of human actions. After all, the covering laws in one branch of human inquiry might be natural laws such as the law of gravitation, whereas in moral philosophy they might be self-imposed rules of obligation. We would still face objections of the kind we considered in Chapter 9, that the covering law model is too narrow for some fields, but implicit determinism would not have to be one of the objections.

Positivism. Positivists are not likely to be determinists. Following the empiricist tradition, positivism will trust as scientific only those lawlike generalizations that are firmly based in experience. But positivists believe with Hume that experience does not yield necessary connections; and without necessary connections among events, determinism would be mere speculation.

Positivists have not always been champions of the freedom position, however. To see why, let us consider the view of A. J. Ayer, who wrote a classic of early twentieth-century positivism, *Language, Truth, and Logic.*[9] Ayer suggested the principle of verifiability as a test of the cognitive significance of any sentence. This principle says that all utterances may be divided into two groups, cognitive and noncognitive. Cognitive utterances are propositions, and they purport to be true—to make a statement of fact. The exclamation "Oh, no!" is not a proposition. It does not purport to be true. Rather, it is an expression of emotion. By contrast, "It is raining" implicitly claims to be a true report of a feature of the world. Now, Ayer said, there are many alleged propositions, such as "God is good," or "You are morally obligated to help persons in need," that turn out not to be propositions at all. How can we tell? The principle of verifiability tells us that an utterance is a proposition only if we can specify the empirical evidence—the sense experiences—that would count as verifying its truth. If we cannot specify such truth-certifying grounds in sense experience, then the utterance may have emotive significance (it may serve as an expression of emotion), but it has no *cognitive* or truth-telling significance.

Armed with the principle of verifiability, Ayer concluded that moral language lacks cognitive significance. Implicit in this position is the conclusion that as emotive expressions, moral expressions do not exist as judgments (propositions) at all. Therefore, they do not force us, as Kant thought they did, to postulate the existence of free selves as either holding rights or standing under obligations.

The emotivist theory of ethics[10] has been much attacked and has been defended in later works,[11] and the principle of verifiability has itself been the subject of much debate and reformulation.[12] Not every positivist need hold Ayer's position, of course, but those who do may use the position to declare both moral language and deterministic theories to be of no cognitive significance and to avoid commitment to either human freedom or determinism.

Pragmatism. It seemed to William James that human choice was an evident fact of life. I can choose which way to walk home, and if someone dares predict which way I'll choose, I just may get stubborn and go another way. While we can imagine the "But I could predict your stubbornness too" response of determinists, James would soon put the question in terms of explanatory payoffs. "Why," he might say, "are you concerned to insist on the possibility of prediction? Because it is a methodological postulate needed for the pursuit of social science? Then, by all means, assume it, but don't trouble us with the demand that we admit the postulate as a somehow revealed truth."

Pragmatism as described in Chapter 7 would be skeptical particularly of those solutions to the freedom-determinism question that demand that one be real, while the other is only an illusion. The favored solution would be whichever one made the fewest metaphysical commitments, while allowing answers to practical problems. Of the positions we have suggested, James would no doubt have favored the one that sees both sides as having appropriate claims methodologically and that refuses to pit them against each other. This conciliatory position has obvious strengths. Its weaknesses have been considered already, earlier in this chapter and in Chapter 7.

The Future of the Question

Debates about freedom and determinism have lately been somewhat out of fashion in work on the philosophy of science. One reason for this neglect is the fear among some theorists that our classical way of asking the question may itself have made any adequate answer impossible. With the advent of the indeterminacy principle in quantum mechanics and the increasing use of cybernetic models as opposed to physicalistic and deterministic models in biology and the social sciences,[13] some philosophers of science believe that determinism is not a position to be taken seriously anymore. Still others believe that there is no way to characterize freedom of the human will since empirical analysis and manipulation seem to be impossible. Finally, increasing specialization in branches of philosophy means that there are fewer thinkers who now take their task, as Kant and other earlier philosophers took theirs, to find an integrated view of all forms of knowledge and their relations with one another.

No matter how we may ignore the problems involved in this issue, however, they are not about to go away. In the matter of the dependence of quantum mechanics on indeterminacy, for example, some scientists and philosophers of science demand a more deterministic alternative.[14] And we find ourselves using deterministic assumptions at one place and free will assumptions at another. Thus on both practical and theoretical grounds there seems to be reason to consider how scientific explanation fits with either freedom or determinism.

Thus far in our discussion of human freedom we have not gone beyond moral choice or choosing to reason. But any question about ethics or the nature of humans cannot long be analyzed without coming to religious foundations for ethics or the relationship of humans to their God, or Creator, or the Transcendent, however it is named. In this chapter we have deliberately left out any discussion of whether humans are or are not related to God and if so, how. Chief among our reasons is the conviction that science and religion have a complex history of interactions and the question of human freedom deserves consideration separately from the larger question. In Chapter 11 we will turn to the question of how religious explanation and scientific explanation fit together, if indeed they can.

Supplementary Reading

Morgenbesser, Sidney, and Walsh, James, eds. *Free Will.* Englewood Cliffs, N.J.: Prentice-Hall, 1962.

This anthology includes writings of Aristotle, Augustine, Hobbes, John Stuart Mill, Sartre, and others and provides the outline for the classical determinist-libertarian debate. Brief editorial comments on the articles keep things in perspective.

Notes

1. Immanuel Kant, *Foundations of the Metaphysics of Morals*, trans. L. W. Beck (Indianapolis: Bobbs-Merrill, 1959).

2. Aristotle, *On the Soul*, in *The Basic Works of Aristotle*, trans. R. McKeon (New York: Random House, 1941).

3. The best-known classical statement of this view can be found in *Leviathan* by Thomas Hobbes.

4. This position that freedom is the opposite of external compulsion and is therefore consistent with fatalism has a long history. It was argued by Hobbes and by John Locke (who was not, however, an unequivocal fatalist). In modern times, the position has been defended by a number of philosophers. For example, see California Associates in Philosophy, "The Freedom of the Will," in *Knowledge and Society* (New York: Appleton-Century Company, 1938).

5. Notice that this analysis seems to equate "predictability in principle" with determinism. If one is a determinist, might it not still be possible to deny that complete and absolutely accurate prediction is possible, even in principle? Yes. We might hold that the world is strictly determined but that we, with our limited powers of observation, can never *show* it to be so. If we take this position, though, we do abandon the claim, being considered in this part of the discussion, that determinism is shown to be true because of scientific evidence.

6. Benedict de Spinoza, *Ethics*, in *Spinoza: Selections*, ed. John Wild (New York: Scribner's, 1958).

7. Such a position is close to that of a major school of Christian thought stemming from St. Augustine and found in Leibniz, among philosophers, and Calvin, among theologians.

8. See, for example, Pierre Teilard de Chardin, *The Phenomenon of Man* (New York: Harpers, 1959); and Theodosius Dobzhansky, *The Biology of Ultimate Concern* (New York: New American Library, 1967).

9. A. J. Ayer, *Language, Truth, and Logic*, 2d ed. (1946; reprint, New York: Dover Books, 1952).

10. More precisely, of metaethics—the field that concerns itself not directly with what is good or right but with the meaning and significance of moral language.

11. For example, C. L. Stevenson, *Ethics and Language* (New Haven, Conn.: Yale University Press, 1943).

12. See Carl G. Hempel, "Problems and Changes in the Empiricist Criterion of Meaning," in L. Linsky, *Semantics and the Philosophy of Language* (Urbana: University of Illinois Press, 1952).

13. For a discussion of these developments, see Jean Piaget, *Biology and Knowledge* (Chicago: University of Chicago Press, 1971).

14. For example, see David Bohm, *Causality and Chance in Modern Physics* (Princeton, N.J.: Van Nostrand, 1957).

/11/ RELIGIOUS EXPLANATION AND SCIENTIFIC EXPLANATION

Science and Religion

It may seem odd to include a discussion of religion in a text on scientific explanation. As a matter of fact, the National Academy of Sciences, at its meeting in October 1972, tried to put some distance between the study of science and religious commitments. The academy was responding to a move in California to tell schools how they should teach biology. The academy adopted the following resolution:

> Whereas we understand that the California State Board of Education is considering a requirement that textbooks for use in the public schools give parallel treatment for the theory of evolution and to belief in special creation; and
>
> whereas the essential procedural foundations of science exclude appeal to supernatural causes as a concept not susceptible to validation by objective criteria; and
>
> whereas religion and science are therefore separate and mutually exclusive realms of human thought whose presentation in the same context leads to misunderstanding of both scientific theory and religious beliefs . . .
>
> we . . . urge that text books of the sciences, utilized in the public schools of the nation, be limited to the exposition of scientific matter.[1]

The academy thus affirms the view that natural science and religion have little or nothing to do with each other. But that is only one view of the matter. We saw in Chapter 10 that several different positions could be defended in terms of the freedom-determinism question. This should lead you to expect the relationship of science and religion to be complex also. And of course the barest knowledge of history will include something of the controversies that have surrounded scientists such as Copernicus, Galileo, and Darwin. Any quick and simple answer to the question of how science and religion are related is bound to ignore many issues of real importance to both scientists and religious thinkers.

All of us make claims and take positions on a wide variety of religious matters. As citizens, we are called upon to judge the extent to which religion should influence the curriculum of public schools, and we all make affirmations of one kind or another regarding the role of religion in our lives. Even the person who says religion means nothing is taking a position on the role of religion and religious explanation.

In this chapter we will explore some of the main views about the connection, or lack of connection, between religious and scientific explanation. What we will call the "confrontation thesis" holds that religious and scientific explanation directly conflict and are incompatible with each other, so that the individual must choose one or the other. In contrast, the "exclusivity thesis" holds that religion and science are such different fields, with different subject matters, and different methodologies, that they are mutually irrelevant. The exclusivity thesis, in other words, seems close to the position adopted by the National Academy of Science, as quoted above.

The title of this chapter springs from a clarification that will help considerably in our analysis of the science-religion relationship. A scientist may love science or view with awe the beauty of the relationships discovered, but the center of the scientific enterprise is the devising of scientific explanation. In religion the human responses of love and awe are more important and lead to worship, celebration, prayer, and exhortation. What we wish to explore, however, is that part of religion that can be called religious explanation and how it is related to scientific explanation. In both types of explanation we are dealing with propositions, that is, statements making truth claims. We will confine ourselves in this discussion to the features of religion having to do with explanation.

One final introductory note: the following discussion is intended to be broadly inclusive of most types of world religions although the examples used will generally be taken from the Judeo-Christian tradition, the one with which the authors are most familiar.

The Nature of Religious Explanation

Religious explanations generally have as their object the answer to a number of questions, the most important of which are the following:

1. Who are we? Or, what is our relation to the Divine or to a source of fundamental meaning?

193

2. How did we get here? Or, where did we as humans come from, and how are we related to the rest of nature?

3. Where are we going? Or, what is our final end, and are we and all of the universe serving an ultimate purpose or destiny?

4. What ought we to do? Or, how should we act toward fellow humans and nature around us?

To illustrate the dimensions of religious explanation, we might refer to the early books of the Hebrew Bible for an account of who the Israelites are and how they came to be identified as the chosen people. The book of Genesis tells how man and woman came to be and how we are to view the rest of creation. It also describes how God made a covenant with Abraham and how the children of Israel, Abraham's descendants, are to conduct their lives. The Hebrew prophets warn of destruction upon disobedience of the Law of Moses and promise Jehovah's favor on those who follow the Law. Modern Judaism and Christianity provide more complex answers to the four basic religious questions because of a belief that the millennia of human history are part of the answer to who we are and how we got here. But the questions are the same.

Another set of answers to the four questions might run as follows: we are self-conscious, thinking beings, and we do not know from whence we have come, but we find no basis for a supernatural source. As humans we owe each other respect, and the highest good is seeking out and developing our human potential to the fullest. The great leaders, scientists, and poets of our civilization give us inspiration. Being true to ourselves is our purpose in life.

Answering these questions, religions will refer to acts of a God or acts of humans in relation to a God. Most religions or organized religious groups will reject answers that are based on the blind mechanical workings of a fate or that assert there is no meaning beyond that which is produced by humans. Thus a religion such as Christianity will oppose an answer that claims I have no more purpose in being here than the rock that landed in my garden during the last meteor shower. The question of purpose, how I got here and where I'm going, is essentially a religious question, and this chapter will explore how answers to such questions relate to answers to questions about the structure of my DNA and how I could fall out a window like any falling body.

A First Position: The Confrontation Thesis

In 1925, a well-remembered trial was held in Dayton, Tennessee. John T. Scopes, a high school biology teacher, was on trial for violating the "antievolution" law of the state of Tennessee. The prosecution (led by William Jennings Bryan—once U.S. Secretary of State and a presidential candidate) argued that the biblical account of man's creation was true (because it was divinely revealed) and sufficient, and that the theory of evolution was in direct conflict with it. The defense, conducted by the famous Clarence Darrow, claimed that the biblical account had to be either metaphorical and therefore possibly consistent with evolution or false since taken literally it was self-contradictory. Scopes lost and was fined $100. Until recently, the law remained on the books.

It has been said that although the prosecutors did not see the strength of the case for evolution, they were correct in seeing the threat that modern science posed, and still poses, for religion. From this view, the systems of religious and scientific explanation predict, or reconstruct, facts that flatly contradict each other. Since rational people cannot accept a self-contradictory worldview, they must choose *either* religion *or* science; these two systems of explanation are at war.

This is the confrontation position, and in this section we shall consider the grounds for it. In the next section we will consider the possible arguments for the position that there is no conflict, or that the conflict between science and religion is the result of a mistake, a confusion of different fields with each other (exclusivity thesis).

Those who see conflict between evolution and creationist accounts of the origin of man generally see this conflict as one that must be won. The adversaries are not happy with a standoff for they do not believe that productive tension can result from coexistence of both views. Propositions about the nature of humans are either true or false, and actions must be decided on. It is either evolution or creation, not both.

Let us examine first the view of those creationists who base their view of human origins on a literal interpretation of the Bible. They believe that in the beginning of time, not too many thousands of years ago, God created the universe and all of the different kinds of plants and animals on this planet. The crown of this creation was humankind or, more specifically, Adam and Eve, who had a special relationship to their creator. Over time some biological species have become extinct and have been deposited as fossils in geological strata by the flood that Noah escaped or by other geological catastrophes. Perhaps the key objection of these creationists to evolutionary theory is based on their conviction that evolution denies any transcendent meaning for human existence. How could we characterize human beings as different *in kind* from animals, as free beings who are responsible for their moral acts, under God, when human beings are held to have evolved gradually and only by degrees from beings who are not held responsible and who are not thought to be anything more than physiological automatons?

There are many creationists (perhaps most?) whose version of the creation account will differ from what has been described, but we will not explore variations here. Instead we will look at another group that sees science and religion irrevocably in conflict on this question of human origins. These are the evolution scientists, who believe that Darwin and biologists since his time have demonstrated the truth about how humans have evolved and what this means. These scientists believe that the earth is about 4.5 billion years old and that life evolved out of a soup of organic chemicals. The characteristic feature of life is the ability to reproduce, and this reproduction occurs with sufficient variation so that from simple single-celled organisms there evolved all the life forms including mammals and finally Homo sapiens about one to two million years ago. Man is thus seen as one biological species among many, each occupying its niche in the environment, and since neither genotype nor environment is holding constant, further evolution will occur in directions we cannot predict. The key feature of this position is the denial that humankind was created by a divine creator and is bound by worship and moral obligation to this divinity. Religious meanings are considered to be superstitions entirely at odds with the scientific facts.

The vigor of the debate between these two groups of confrontationists is matched by the splendid variety of views that are held by individuals who have written in this field. This brief account of the two positions is pale in relation to the richness of the views and the sweep of the rhetoric employed. Our purpose cannot be to explore the variety of positions but is rather to examine whether the conflict between science and religion has relevance for the question of what constitutes a scientific explanation. Both groups seem to be saying something like this: scientists discover facts about our universe and these facts are incompatible with the facts of human origins given in Genesis. If the scientists' facts are correct, then Genesis and the Bible are wrong and man is an animal without moral obligation to a creator.

Note the character of scientific fact in the hands of the confrontationist. Facts are considered to be immutable or unchangeable, and some of these facts are known with certainty. This is a view of scientific knowledge that has no room for tentativeness and seems incompatible with the common idea that scientists are antidogmatic. Certain knowledge is surely not what a positivist would claim, nor is it consistent with Copi's position on crucial experiments. The confrontationists seem to require a correspondence theory of truth (see Chapter 3) along with the conviction that some of their scientific propositions have been shown to match the physical universe accurately.

One group of creationists, who call their position scientific creationism, claim that scientific evidence supports the creationist position; but they also hold that creationism is beyond refutation by observation and experimentation. This is hardly consistent with any normal use of the term *scientific*, and we will not describe the position in that way as it can only result in confusion.

Another characteristic of the confrontationists is their apparent acceptance of a reductionist position. If humans are to be understood simply as bags of chemicals arranged in a certain way by the chance interaction of genes and environment, then there would appear to be no possibility for transcendent meaning in human existence. This is a reductionism that leaves room only for materialism and implies a commitment for scientific explanation about the nature of reality that we discussed in Chapter 6. But the creationists are equally reductionist in their emphasis on the spiritual. Not even a Cartesian dualism is possible if physical objects and the laws relating those objects are so inconsequential that they can pass into or out of existence at the whim or word of a deity.

However, the acceptance and use of the theory of evolution in biology does not necessarily lead us to so rigid a confrontationist position. We have examined a number of theories that would allow alternatives to reductionism. Among these are the following:

1. We have seen (particularly in chapters 3 and 10) that Immanuel Kant suggested making a sharp distinction between appearance and reality. To science

he assigned the task of organizing the appearances while taking care that it did not make judgments about the world of things in themselves, behind the appearances. Seen in this light, evolution is a theory that organizes empirical knowledge, but it would certainly not give us justification to conclude what human reality is ultimately like.

2. Theodosious Dobzshansky[2] developed for biology the notion of emergentism, which has also been defended by Teilhard de Chardin. The emergentist thesis (discussed in Chapter 10) holds that although evolution proceeded originally from inanimate matter, eventually mind developed and took on a reality and qualities of its own, no longer reducible to the material from which it came. This provides another way to affirm the theory of evolution but deny that reductionist consequences flow from it.

3. We have noted that positivists and pragmatists are conservative in the conclusions they are willing to draw from scientific theories involving laws. They are skeptical because they see these laws as merely generalizations, for purposes of economy, from past observations and reports. The positivist skepticism flows from the view that these laws possess no certain applicability to the future and that they are therefore not generalized descriptions of the way reality has to be. It is clear that in either a positivist or pragmatist view, reductionism is never a valid *conclusion* from past data.

Just as the theory of evolution is subject to a number of interpretations, so also may the Genesis account be read in different ways. First, that account may be symbolic. Or it may be taken to be literal and fallible. Religious explanations, typified in this case by the Genesis account of creation, if taken in a way other than literally, may be held to refer to the meaning of events rather than to scientific facts or appearances. Thus interpreted, some religious explanations at least may be held not to make factual claims about appearances at all, and therefore they could hardly be held to be competitive with scientific claims. They could, for example, carry meaning for humans as free and responsible beings.

This much is clear: many creationists and many evolutionists see little reason for confrontation. They find ways to frame their views on creation and evolution that are not at all confrontational.

We have seen that the confrontation thesis requires some assumptions about the nature of science and religion. When scientific explanation is found to conflict with religious explanation, it is because certain philosophical assumptions have been made. Further, there are several ways in which the conflict can be attenuated. This raises the possibility that perhaps these two accounts aren't talking about the same thing.

A Second Position: The Exclusivity Thesis

According to the exclusivity thesis, scientific explanation and religious explanation are of such a different character from each other that they are mutually irrelevant. On the surface, this seems an unlikely thesis because they both appear to deal with the same subject, human action; moreover, the word *explanation* is used in common. In order to understand the different bases for the exclusivity thesis, we will therefore need to investigate the fundamental distinctions that allow us to put these two forms of explanation in such different categories. In this section we will examine three ways of making these distinctions, any of which might be used as a foundation for the exclusivity thesis. We will find that the three paths are ones with which we are already familiar because they have figured importantly in characterizing scientific explanation itself and in relating it to ethics.

Distinction 1: Fact Versus Emotion. If one could establish that religious explanation was actually not real explanation at all but only an expression of emotional reaction to the world and our place in it, one could then hold that scientific and religious explanation could not compete, because only the scientific is genuine. A position close to this was developed by A. J. Ayer, in *Language, Truth, and Logic.*[3] We have seen this positivistic position developed earlier (in Chapter 4 and again in Chapter 10) and will therefore only briefly restate it here. Positivism holds that all knowledge comes from sense experience and that a proposition can be called cognitively meaningful (that is, meaningful in the sense of actually asserting something) only if you can show what experiences would confirm it. Thus, all cognitive propositions or statements of fact are empirical and are the province of science. If you assert that a religious explanation is *the* correct, or *a* correct, account of an event, you must be prepared to submit that claim to scientific scrutiny. No private appeal to the word of God for a faithful community or individual intuition or inner experience will help. Moreover, no statements about what ought to be are directly provable by reference to what is, to facts. Therefore, any value words or phrases in the explanation will be giveaways: such assertions will not be cognitively meaningful. To say, for example, "God loves us," looks like a statement of fact. For A. J. Ayer, this is not a statement of fact because the speaker is not likely to be able to cite experimental data to support the claim. It must therefore not *mean* what we thought it did. It's likely, said Ayer, that we mean something like "we ought to live a certain way or have a certain attitude toward life, acting as if a benevolent father were running the world." Taken this way, the expression is more like a positive attitude toward life in the world. It is not an explanation or a statement of fact at all.

As we have seen in earlier discussion, the great power of the positivist thesis is its attempt to tie scientific claims to empirical evidence and its attempt to keep pseudoclaims out of competition for scientific attention. It obviously fits with a claim to the complete separateness of science and religion. We find, however, that there are a number of difficulties with positivism as it is applied to the question of religious explanation by Ayer.

1.   The clearest cases of noncognitive utterances are sentences like "Boo!" or "Ouch." These expressions reveal attitudes, without making factual claims. But it strains the sense of the meaning of religious explanations to say that they are no more than expressions of emotion, such as "Ouch." For one thing, reli-

gious explanations seem really to be about the world we live in, and they are intended to make a difference in the way people look at our world. They are certainly not intended by those who offer them to be irrelevant to the facts of our lived existence.

2.   Ayer's position assumes that it is possible to make a rigid distinction between facts and nonfacts. It therefore must ignore or discount the extent to which methodological assumptions, desired results, and other theoretical commitments figure into the way we see facts and organize them. We have seen (in Chapter 4) that it is hard to make the case that human beings can see the empirical facts "as they really are," without assumptions and values being involved.

3.   Religious explanations present difficulties when we try to reduce them to an expression of emotion. In fact, such explanations are sometimes as contrary to our emotional states and expectations as are facts of science. It may be a fact that a hurricane is coming, even though we wish it weren't. And religious literature is filled with the same kinds of explanations contrary to our wishes. Sometimes people recognize a "call" to a given profession or a vocation, even though they would rather do something else. The Old Testament prophets commonly cited explanations for events that did not sit well emotionally with those who heard and accepted them.

Of course a person who wants to defend the position at all costs may argue around the difficulties mentioned above by saying that religious explanations are not what they appear to be. But the defense of a position in this way is always a drastic step because it is difficult to know when one is getting at the "real meaning" and when one is simply ignoring relevant data. It is particularly embarrassing for the positivist position to ignore the apparent and intended meanings of utterances when the position itself celebrates the virtue of looking at experience without constructing a hidden reality behind it.

Positivism has not just been attractive to some scientists. Those interested in religious language have sometimes felt attracted by it because it does seem to insulate the meaning of religious expressions from scientific criticism. Put another way, and in language with which we became familiar in Chapter 10, the position is a way to defend the autonomy of value statements. That is, it would argue against the *reducibility* of religious utterances to scientific ones. Let us examine a position that, like positivism, preserves the separation of facts and values, without making Ayer's move of calling religious utterances merely expressions of emotion.

Distinction 2: Values Versus Facts. We might propose that scientific and religious explanations are not relevant to each other because science deals with facts and religion deals with values. Are values and facts separate? In an important sense, they surely are. When you say, "I wish it would rain," you're not confusing this value with a fact because you know it isn't raining. In general, when we wish that something were so, or say we want it to be so, we are valuing *contrary* to fact. Even when we say that something is good, we mean far more by

that than merely that it exists. We can describe facts, or psychological properties, or in general the way the world is, and we can also pronounce on the value
of this creation, of this world we live in, but we ought never to confuse those
two acts of description and valuing.[4] The fact that something is so never auto-
matically also makes it good. Recall that in the Genesis account of creation,
God created the world, and then, apparently in a separate act, "saw that it
was good."

Does this approach solve our problem and show how religion is different
from science? If so, we can count on religion to investigate and pronounce on
the goodness of different aspects of creation or human action, while we let both
the human and natural sciences go about their own business of describing the
way things really are. Before we agree too quickly, however, let's consider the
assumptions on which such a position might rest.

One of the assumptions underlying our new possibility is that we can really
separate facts and values, that it is possible for us "merely to describe," and it is
also possible for us "simply to value, without reference to facts." But Irving
Copi, in his essay "Crucial Experiments" (cited in Chapter 4), shows us some
profound problems with the view that pure description is possible. In order to
describe what is happening with some confidence, we need to affirm a large
number of assumptions about other variables. The trouble is that we never seem
to be able to reach those simple descriptions of the world that don't depend on
any other assumptions, and therefore no matter how clean of values we try to
keep our descriptions, we are always smuggling in theories in which we have
more or less confidence or which we value for their power. Values are involved
every time we read a thermometer because we round off the reading to the
nearest degree or so. We say that we have rounded it to a certain number of
significant figures, and in this case we mean that we don't care about, we don't
value more precision than that. Pure description, without any values assumed at
all, may be an illusion.

Pure valuing is at least as difficult as pure description. When we value
something, remember that we are valuing *some thing*. Our values include references to the world. When we say we want something to happen, we are making
a host of claims about what is possible, about the consequences of the occurrence for which we hope, and so on. This fact has led many thinkers to believe
that the foundation of all ethical theories is in religious affirmations. They
suggest that to take a stand on the question of what we ought to do or ought to
be is always to assume a set of judgments about the way the world is and its role
as a product of divine creation.

What we have just said muddies the water that we hoped a sharp fact-value
distinction would make clear. The arguments suggest that a science that claims
to deal only with the facts is going to smuggle its values in one way or another
anyway and that the valuing that characterizes ethical judgments and religious
affirmations will have inevitable reference to what we think are facts about
the world.

Is there any way to separate religious and scientific explanations without
having to face the foregoing difficulties? A third distinction might help us to do
this, by trying to separate the very languages that religion and science speak.

Distinction 3: Appearance Versus Reality. We have seen in previous chapters that Immanuel Kant tried to distinguish sharply between the world of things as they appear to us in experience and the world as it really is, independent of human perception. Kant believed, and many thinkers since have agreed, that science found its proper role in dealing with the realm of appearance. Science should not attempt to tell us about the world of things in themselves because no sensory evidence was available or could ever be available to gain access to that realm. When we make a claim, we always look for evidence *within* our experience. When we try to peer beyond our experience, we face a contradiction. The very "peering" means we're looking for more sensory evidence and we find ourselves confined once again only to what experience can teach us.

Yet as we have seen, Kant believed that it is necessary for us to act as moral agents in the world. Indeed, who can dispute it? We face choices every day, choices that force us to envision the world as it ought to be rather than as it simply is at the moment. But this vision of the world as it ought to be is by definition *not* the world of human experience. We are led to say that something ought to be different just because we want to go beyond the world as we find it.

It appeared to Kant that if we feel ourselves under obligation to do what is morally right, as all of us do (even when we don't obey the strictures of our conscience), we are forced to say a number of things about the world as it really is. These are things that we would never as scientists dealing with experiences be justified in saying on the basis of sensory evidence. Among these affirmations that we feel forced upon us are that human beings are free, that our souls are immortal, and that there is a divine judge who ultimately governs the world in such a way that those who deserve happiness are given it.

We will not consider the argument Kant used to try to establish these claims about the world as it really is, beyond human experience. They are presented succinctly in Kant's *The Foundations of the Metaphysics of Morals.*[5] We can see, however, the power of the analysis in solving some of the problems the earlier distinctions have presented us. We can hold that both scientific explanations and religious explanations are *cognitive*; that is, they are capable of being true or false and capable of having good reasons given for why we ought to believe that they are true or false. And yet religion and science need not compete because they are dealing with entirely different realms—appearance, in the case of science, and reality, in the case of religion and ethics.

As always, unfortunately, there are costs for the advantage Kant gained for us. Beyond the bare claims that humans are free and immortal and that God reigns in the universe, there can be no science of theology. In order to account for the fact of human moral obligation we can make these two claims and no more. When theologians begin to spin theories about how the world came to be in God's creation or what the destiny of humankind might be, they get no support from Kant. He insisted that all such attempts will be pseudoscientific because they will pretend to adduce evidence when the only possible evidence could come from experience; yet experience is the realm of science and always tells us only what is in our appearing world, never what the truth can be beyond all possible experience. Private devotion, prayer, celebration, and worship are possible for the religious. A public science of the supernatural is not possible: it is an illusion. That may be enough for many religious people. But it is certain to be unsatisfying for those who believe that religion can provide explanations.

William Ernest Hocking, in *Science and the Idea of God* revealed his own dissatisfaction with the Kantian solution.

> For a century and a half since Kant, first ethical then psychological and sociological interpretations of religion have been pressed to provide substitutes for the doctrinal elements of faith. Religious ideas become "postulates." That is to say, demands made on the world, launches of will, wills to believe. Or religion itself is a phenomenon of self-consciousness, a factor of social adjustment, a semi-beneficent illusion: we cannot say, "There is a God," but we can say, "Man is a praying animal"; scare him enough and he prays, revealing something "deep in subconsciousness" and therefore highly authentic. Of such anthropological verity is much contemporary religious thought constructed, science now holding the whip of a resigned authority. Religion shows gratification when the anchor of its drifting boat catches in a submerged tree!
>
> Shall we be satisfied with this situation? To my mind this is less than peace by appeasement; it is peace by capitulation.[6]

We have seen several possible ways to argue that religious explanation and scientific explanation cannot compete because they are irrelevant to each other. We can do so by distinguishing between fact and emotion and giving science the province of fact; we can sharply distinguish between facts and values and again retain for science the realm of facts; we can make a sharp distinction between appearance and reality, and let religion handle reality while science organizes the appearances.

In the introduction to this chapter we quoted from a position statement of the National Academy of Science: "religion and science are therefore separate and mutually exclusive realms of human thought." It seems clear that this is consistent with the exclusivity thesis, but there are few clues as to which of the foregoing three distinctions would be most appropriate. It is often the case that exclusivity is claimed either by scientist or theologian just to get the other out of the way. Whichever distinction lies behind the National Academy's statement, it is clear that each position has attractions and there are adherents for each position. We have also seen, however, that each position has its difficulties. Therefore, some have tried to avoid both confrontation and exclusivity by embracing both scientific and religious explanations.

A Third Position: The Double-Aspect Theory

The double-aspect theory attempts to make the best of what we have seen is a bad situation. The various attempts to show that scientific and religious explanations either always compete with each other or are irrelevant to each other have serious difficulties. A central thread in these difficulties is the problem we have in isolating just what the facts of our experience are, and trying to separate rigidly from them our wishes, our values, and our conceptual frameworks. Double-aspect theorists urge us to recognize these difficulties rather than try to eliminate them. Why don't we just admit that any attempt to describe the facts of our experience will depend upon our making value assumptions and framework assumptions. Let us admit that there are no uninterpreted facts. Perhaps we can say that both religious explanations and scientific explanations are interpretations of, *and abstractions from,* our experience. We can then say that religious knowing and scientific knowing grasp two *aspects* of reality but that neither can claim to include all of the richness that the other can lay claim to. Such a theory suggests that we can no longer distinguish between scientific theories and the most important religious ideas. Just as religions tell stories to try to illustrate the world as it appears from a religious perspective, so we can say in parallel fashion that scientific theories are stories or models of their own kind. Remembering Immanuel Kant's point about the highest-level framework assumptions that underlie scientific knowing, we might say that the claim that all events in the world are connected as causes and effects to one another is comparable to a religious claim that the world is the product of intelligent design. Both of these high-level claims are not facts in any ordinary sense. They are rather the framework within which facts are recognized and related.

Notice that one advantage of the double-aspect theory is that it does not force us to make the radical distinction between appearances and things in themselves (or the world beyond all human experiencing) on which Kant relied. It tells us that even in this world of sense experience there are no "hard data" that can ultimately decide between competing frameworks. The hard part of this doctrine is that there will be no single adequate interpretation of the world because all interpretations are evaluationally charged and include within themselves the criteria by which all more specific claims are to be judged. However, this need not lead to a kind of vicious relativism that would claim that there is no way to decide between competing frameworks. We might compare a religious framework of explanations that maximizes our attention to the transcendence of the spirit over the material world and our relation to the divine with a scientific explanation that attempts to maximize predictability, etc. We might use both and simply recognize that our purpose in explaining is different at different times and for different ends.

The difficulty with the double-aspect theory is the same as we encountered in Chapter 6 when we saw it as a way to explain the relationship between mind and body. It seems to solve many of the problems by insisting that one can choose neither side. But the hard questions of what I am to believe at a particular time are still there. When I face increasingly sophisticated psychological, sociological, and biological explanations for why I make a particular choice given two alternatives, how can I any longer claim to have made a moral choice in response to divine leading. In other words, the *unity* of our world is not accomplished by the double-aspect theory. It counsels us to be content with an irreducibly many-sided account of what the world is like, and it offers no firm guidance when these various sides come into explanatory conflict with one another.

A Return to Theories of Scientific Explanation

This chapter has considered a variety of positions on the relation of scientific and religious explanations, but we have seen throughout that the standard accounts of what it means for an explanation to be scientific and how it should best be constructed have found their way into each of these points of view. A few examples may help make this clear.

Positivism can be used not only as a ground for understanding that scientific explanations are in fact descriptions (Chapter 4); it can also provide the basis for excluding religious explanations as legitimate as public accounts because they can be held to be noncognitive, nonfactual. At the same time, some defenders of religious explanation may be attracted by positivism simply because it makes the demarcation between the religious and scientific spheres so sharp.

Kant also sharply distinguished the languages of religion and science, but he did so by seeing religious language as justified by our moral obligations and as having nothing essentially to do with reports of our experience. The distinction allows for the autonomy of religion but purchases that autonomy at a great price, for it cannot allow religious language to be publicly meaningful.

Pragmatism seems close to the final position that we have considered in this chapter, the double-aspect theory. Pragmatists would insist that ultimately the choice of an explanatory framework depends on what we need to know, to understand, and even to identify with.

We have arrived at no answers in this chapter. Yet a great deal has been accomplished if we can use our acquaintance with the various theories that have been considered in this book to realize that simple resolutions of the relation between religious and scientific explanations will be hard to come by. Those thinkers who most clearly demarcate various explanations as clearly competing or clearly irrelevant to one another can be seen to have done so only at the cost of a large number of assumptions that we have found cause to question. In our examination of various theories of scientific explanation we now have seen how scientific explanation fares when brought next to religious explanation. In Chapter 12 we will review where our search for the best theory of scientific explanation has taken us.

Supplementary Readings

Barbour, Ian. *Issues in Science and Religion.* New York: Harper & Row, 1971.

This is an excellent source for answers to a variety of questions about science and religion. Barbour's treatment of the history of this relationship is exceptional.

Hick, John. *Philosophy of Religion.* 3d ed. Englewood Cliffs, N.J.: Prentice-Hall, 1983.

Hick examines the relationship of science and religion from a Christian perspective and writes well about the truth claims of religious language.

Peacocke, A. R., ed. *The Sciences and Theology in the Twentieth Century*. Notre Dame, Ind.: University of Notre Dame Press, 1981.

This set of papers given at a symposium is substantial and difficult to read, but it is worth the effort for those who want to go into greater depth.

Notes

1. Quoted in William H. Austen, *The Relevance of Natural Science to Theology* (London: Macmillan Press Ltd., 1976), p. 1. The controversy in California continued. In September of 1984 the board of education rejected all seventh and eighth grade texts offered as science texts by American publishers on the grounds that the treatment of evolution was barely visible or inadequate. To avoid creation-evolution debate the publishers had said virtually nothing on the subject.

2. Theodosius Dobzshansky, *The Biology of Ultimate Concern* (New York: New American Library, 1967).

3. A. J. Ayer, *Language, Truth, and Logic*, 2d ed. (1946; reprint, New York: Dover Books, 1952).

4. G. E. Moore, in *Principia Ethica* (London: Cambridge University Press, 1903), developed a detailed and subsequently well-known argument concerning the separateness of facts and values. It is a good starting place for further reading on this issue.

5. Immanuel Kant, *Foundations of the Metaphysics of Morals* (1785; reprint, New York: Liberal Arts Press, 1949).

6. W. E. Hocking, *Science and the Idea of God* (Chapel Hill: University of North Carolina Press, 1944).

/12/ EXPLAINING AND PHILOSOPHIZING

Why Do We Philosophize?

A few years ago, a Supreme Court justice, on being asked to define obscenity, said that he couldn't define it, but he knew it when he saw it. The question reflected the importance of seeking clarity when policy matters of great consequence rest with the meanings of our words; the answer reflected our sense that somehow we do know what we mean without drawing neat little boxes around each concept we use. Both perspectives are worth remembering as we reflect on the meaning of explanation. We have not arrived at a clear and agreed upon meaning for the term, even though we all use it often; and we have seen that its importance makes continued seeking worthwhile.

Why is explaining so central an activity? In one sense, it is pointing to the fabric of the world we live in, the texture, the substance, the phenomena that are all around us. That way of thinking of it finds expression in explanation as description. Or one may think of explaining as seeking the structure that binds the phenomena together. It may be logical "thread," as the covering law model suggests, or the purposeful handiwork of God the creator. And these ways of stating the idea of explaining may be usefully different and complimentary, or opposed and mutually exclusive. We have seen both these possibilities represented, especially in Chapter 11. However one characterizes it, explaining is a regular and important part of our response to the world.

Plants and other animals do not seem to need theories of explanation. Trees grow toward the light, sending their roots in the appropriate direction. Many animals follow patterns learned by imitating adult behavior. Though some animals use what may be very sophisticated language, others get along quite well without it. Why are humans different? Why do we have such a consuming interest in explaining and then even in asking and arguing about what we are doing when we are explaining?

Aristotle provided an answer to the question of human uniqueness that has continuing force: "All men by nature desire to know."[1] We behave, and adapt, and think; but to this is added a further and wonder-producing capability. We are aware of our behaving, of our adaptations or resistance, of our own thinking. When memory is added to this capacity for self-awareness, enough culture to provide the collective memory shared by people across generations, and the desire to know, then explaining can be seen as an essential human activity. It is part of the expression of humanness itself. We are driven to think and worry and argue about it!

The sense of wonder motivates some of us to philosophize. We want to know the lay of the land before traveling too far or for the sheer joy of reading the map. For others, activity and accomplishment of the tasks and challenges at hand are primary. Maps will be consulted by this group not for the pleasure of it but to help get on with the journey. We have encountered thinkers of both dispositions. (William James was thinking of something like this distinction when he contrasted tender-minded and tough-minded philosophers.) Perhaps most of us find something of each type in our own personalities at different times and with respect to different issues. But the point is this. There are many reasons to philosophize, many sources for our concern about such big topics as explaining. Philosophy unites in common inquiry those who are led to it by a sense of wonder and those who are driven to it by a need for the clarification of concepts that will let them get on with their work.

Doing Philosophy and Getting Answers

If we humans seek explanations and ask philosophical questions, when do we get the right answers? Is there an end point for philosophical inquiry, or do we just question forever? And if questioning never ends, what is the use of it? These are ultimate questions, much like the question What is the world really like? We do come up with answers to these questions; though we are likely to return to them again and again because the answers we were comfortable with earlier turn out to be inadequate to some new purpose.

Lessons from the advances and retreats in the history of science can help us keep perspective on the question of progress in philosophy. In both science and philosophy we use theories to represent the world as we see it or as we think it must be to account for what we see. These theories almost never match the evidence in any really clear way. They may lead us to seek new evidence to confirm or falsify them; they may suggest that what we thought were good data earlier really were not. And when in all their strength they still don't measure up to what we see or what they predict, we often modify them before throwing them out. Elegant theories, whether we call them scientific or philosophical, are some of the finest products of our collective human effort. We do not discard them easily, though we try to make them prove their worth. In this effort for greater understanding, which stretches over the centuries, we can be said to advance in our insight and our wisdom when and as we remember and incorporate the strengths of positions; and we remember what previous attempts have left out or deformed or confused as well.

Philosophy, then, like science, can give us a sense of accomplishment even though the task of understanding everything is of course in one sense endless. Therefore, why not now summarize all that we know for sure, write it down, and agree on it as a mark of current progress?

Though textbooks that taught philosophical truths would be appealing in one way, as are compendia of chemical formulae or taxonomies of plants and animals, philosophy, like science, is not a collection or set of something else. It is above all an activity. Plato counseled us that it cannot be taught but that it may be advanced best as a conversation among friends. That is good advice for many reasons. For one, big questions require deep and very personal desire—the kind of straining for the key that we have seen is also characteristic of great scientific thinkers. The truth cannot be passively absorbed. For another, just as an athlete demands more difficult tests as earlier ones are mastered and psychologists may find that earlier therapies don't work with a new patient, we return to our view of the world again and again, alone and in groups, modifying, adapting, and occasionally reorienting.

The dialectical (contrasting points and counterpoints without definitive resolution) argument of this book has been an attempt to remain faithful to the above judgments about the nature of philosophical inquiry. We have intended it as a guide but also as a companion in the reader's attempt to come to philosophical activity personally. Just as a text such as this is always second best to conversation among friends, so also parting advice has the limits of one-sidedness and the risks of answering questions that have not been asked. Mindful of these problems but also wishing for the reader the best chance of a continued and fulfilling inquiry into the fundamental questions of human life, we offer the following counsel:

1. Embrace your passions for knowing and your heartfelt convictions on whatever questions are driving your inquiry. We sometimes think of philosophers as utterly fair-minded (that is, passionless) sages who think and write truth after truth as each is proved. That is a silly model of human inquiry. Philosophers, like scientists, are usually driven to think and write by a conviction that something they think is right. They therefore set out to show everyone else that what they have come up with is both true and useful. Sometimes we succeed in that, and sometimes we end up realizing that we were wrong to have been so sure. But our passion for the truth and even our passionate conviction that we are right about some point can be our greatest ally. It energizes us and makes our inquiry human.

2. Seek alternative points of view as clues to dimensions of a problem you've not yet seen, and try to imagine each time you find a view opposed to your own what powerful reasons the other person had for defending that view. That procedure throws us into a sympathetic community of inquiry in which we can learn best from and with one another.

3.   Recognize that you will always be steering between two dangers as you philosophize. First, in an attempt to be open to the variety of views, you may come to value compromise more highly than coherence. We've seen that the strongest position on the nature of law in explanation or on the freedom of the will may involve weaving a theory from several different strands and points of view. But it can be beguilingly easy to affirm several appealing but contradictory points of view. Then you end up with a list of separate claims that possess no unity and from which you can predict nothing, or anything anytime. Second, in an attempt to save the coherence of your worldview through time, you will be tempted to ignore or diminish the importance of new questions and issues that dealt with openly could challenge you to change your mind on fundamental points. Consistency is a virtue; unchanging views through much of an adult lifetime may be not so much a virtue as a signal that one is no longer paying attention.

4.   Seek, nurture, and cherish communities of philosophical inquiry among friends. The trust and openness to explore fundamental questions of knowledge and life, and the energy to pursue the truth are gifts we can give one another, and they are gifts that affirm our humanity in ways nothing else quite matches.

Note

1.   Aristotle, *Metaphysics*, in *The Basic Works of Aristotle*, trans. R. McKeon (New York: Random House, 1941), p. 689.

TEXT CREDITS

INDEX